SpringerBriefs in Food, Health, and Nutrition

Springer Briefs in Food, Health, and Nutrition present concise summaries of cutting edge research and practical applications across a wide range of topics related to the field of food science.

Editor-in-Chief

Richard W. Hartel, University of Wisconsin—Madison, USA

Associate Editors

J. Peter Clark, Consultant to the Process Industr
John W. Finley, Louisiana State University, US.
David Rodriguez-Lazaro, ITACyL, Spain
David Topping, CSIRO, Australia

For further volumes:
http://www.springer.com/series/10203

Silvana Martini

Sonocrystallization of Fats

Silvana Martini
Nutrition, Dietetics and Food Sciences
Utah State University
Logan, UT
USA

ISBN 978-1-4614-7692-4   ISBN 978-1-4614-7693-1   (eBook)
DOI 10.1007/978-1-4614-7693-1
Springer New York Heidelberg Dordrecht London

Library of Congress Control Number: 2013939049

Printed on acid-free paper

Springer is part of Springer Science+Business Media (www.springer.com)

Contents

Chapter 1
Introduction

1.1 Importance of Fats in Foods

Lipids, proteins, carbohydrates, and water are the major constituents of foods. Lipids impart nutritional and functional properties to foods. From a nutritional standpoint, lipids constitute a high source of energy (9 kcal/g) and are also an excellent source of essential fatty acids such as linoleic (an ω-6 fatty acid) and α-linolenic fatty acids (an ω-3 fatty acid). Lipids also play an important role in the physical properties of foods since they impart specific characteristics demanded by consumers such as texture, mouthfeel, and melting behavior. For example, consumers expect butter to be solid at refrigeration temperature (5 °C), partially melt at room temperature (25 °C), and completely melt at body temperature (37 °C). Another typical example is that of cocoa butter where the melting behavior and texture of this fat play an important role in consumer acceptance and product quality. Cocoa butter has a sharp melting behavior where most of the crystals are melted at 35 °C which results in optimal flavor release, ideal mouthfeel, and a characteristic "cooling sensation" typically found in chocolate products. The characteristic "snap" and "glossy" properties of chocolate are also related to this sharp melting behavior of cocoa butter.

Chemically, edible lipids are complex materials composed mainly of triacyl-glycerols. These molecules are formed by three fatty acids esterified to one glycerol molecule. Fatty acids can be esterified in different positions in the glycerol molecule resulting in different types of triacylglycerols. The fatty acid and triacylglycerol composition of the material determine the physical state of the lipid at room temperature. Lipids with high content of saturated and *trans*-fatty acids are solid at room temperature and are usually referred to as "fats," while lipids with higher content of unsaturated and polyunsaturated fatty acids are liquid at room temperature and are usually referred to as "oils." Oils are most frequently obtained from plants (soybean, canola, olive), while fats can be obtained from animal sources (lard, tallow, milk fat). However, there are exceptions to these rules. Some semi-solid fats can be obtained from certain plants such as palm (palm oil, palm kernel oil), cocoa (cocoa butter), and coconut (coconut oil) and certain liquid oils can be obtained from animal sources such as fish or crustaceans.

The heterogeneous chemical composition of edible lipids is responsible for a broad range of properties found in fats and oils including texture, solid/liquid ratio, and melting behavior. The diversity of functional properties found in lipids is the basis for their use in different food applications, from moisture resistance agents in baked items to emulsifiers, creaming and lubricating agents (O'Brien 2009).

1.2 Fats and Processing

Animal fats were the first lipids used for food production. They contain a significant amount of saturated fatty acids and impart appropriate structure to foods. However, the use of animal fats in food products has been limited given nutritional concerns about the high saturated fatty acid and cholesterol content, and high costs associated with their production. Gradually, the North American food industry has replaced animal fats with vegetable oils. The lack of texture and plasticity of oils with high contents of unsaturated fatty acids and low contents of saturated fatty acids has limited their use in many food products that require a highly structured lipid. To overcome this limitation, processing technologies and new varieties of vegetable oils have been developed to expand the functional properties of fats and oils. Some of these technologies include: hydrogenation, fractionation, blending of different oils or oil fractions, and interesterification. These technologies offer the capability to produce lipids with different solid/liquid ratios and various functional properties including consistency, firmness, mouth-feel, flavor, crystal habit, appearance, nutrition, and spreadability (O'Brien 2000).

Partial hydrogenation has been a common processing technology used to improve the functional properties of vegetable oils. Depending on the degree of hydrogenation, different solid/liquid ratios can be tailored for specific applications (O'Brien 2000). However, recent research has shown a correlation between *trans*-isomers generated by partial hydrogenation and the incidence of coronary heart disease (Mensink and Plat 2008). This has resulted in the gradual elimination of *trans*-fats from most food formulations. Food producers have been challenged to find replacements for *trans*-fats with optimal functional properties and improved nutritional profiles. Palm-based oils have been the main source of *trans*-fat replacers since these semi-solid vegetable oils have appropriate functional properties and provide adequate physical and sensory properties to foods. However, the high saturated fat content in these shortenings is still a concern since high saturated diets are also correlated to cardiovascular disease (Mensink et al. 2003; Grande et al. 1970; Bonanome and Grundy 1988). Other approaches adopted by the food industry include the use of high stearic shortenings produced by either full hydrogenation, plant breeding, and/or interesterification. The use of these fats with a high content of stearic acids has interested the food industry as stearic acid appears neutral toward the incidence of cardiovascular disease

(Mensink et al. 2003). The main disadvantage of these shortenings is that their high stearic acid content leads to a hard and waxy crystal network that makes them inappropriate in many food formulations. One strategy currently used by the food industry is to blend different lipid sources such as palm-based lipids, high stearic fats, and low saturated fats, to tailor a fat for specific applications. This enables producers to balance nutritional and functional properties of shortening. The ideal solution to the *trans*-fat replacement problem is to use a lipid source that is low in saturated fatty acids and has the appropriate functional properties required by the food industry. The food industry has successfully eliminated *trans*-fatty acids from many of their formulations; however, there is still tremendous opportunity for researchers to find alternative lipid sources and/ or novel processing conditions that can improve the nutritional and functional properties of shortenings and other functional fats.

The functional properties of fats are determined by the chemical composition of lipids. During processing of a lipid-based food, a cooling step is usually included to induce some degree of crystallization in the fat. The characteristics of the crystalline network formed depend on the fat's chemical composition and on processing conditions. Cooling rate, crystallization temperature, the addition of emulsifiers, and agitation are some of the processing conditions that influence the functional and physical characteristics of the crystalline network formed. Therefore, shortenings' functional properties can be tailored for specific applications by (1) changing the chemical composition of the lipid and (2) changing the crystallization behavior of the fat (Martini et al. 2006; Peyronel et al. 2010). Changes in the crystallization behavior of lipids can be achieved by optimizing processing conditions or by using novel technologies that can modify the crystallization behavior of the lipid.

1.3 Sonocrystallization

Sonocrystallization has been used for several decades in aqueous systems but the use of this technique in lipids is relatively new. The interest in sonocrystallization of fats started in the late 1990s and early 2000s when several research groups used ultrasound to change the crystallization kinetics of edible lipids (Higaki et al. 2001; Patrick et al. 2004; Ueno et al. 2003). Specific applications of ultrasound have been described for use in the confectionery industry where patents have been published explaining the use of ultrasound to induce the formation of a stable polymorphic form in cocoa butter (Arends et al. 2003; Baxter et al. 1997a, b) and delay blooming in chocolate. Based on the existing data, our laboratory has used sonocrystallization in fats to modify their crystallization behavior and also to change their functional properties such as texture, viscoelasticity, and melting profile (Martini et al. 2008, 2012; Suzuki et al. 2010; Ye et al. 2011). The ultimate goal of this area of research is to use sonocrystallization as a complementary processing tool to induce the crystallization of lipids and to generate materials with improved functional properties. Research in this field is still in its infancy with much yet to be explored. Some of these unexplored areas will be discussed in detail throughout the chapters of this Brief and include topics such as the effect of processing conditions and nature of the starting material in sonocrystallization processes. More research is needed in the area of sonocrystallization of fats to better understand the conditions needed to tailor fats for specific food applications using sonocrystallization.

The sonocrystallization of materials is achieved using a specific type of acoustic techniques. Acoustic techniques can be classified into diagnostic, high frequency, and power ultrasound based on frequency and power of operation. Non-invasive techniques such as high frequency ultrasound are used as analytical tools to monitor physical or chemical changes in materials. More invasive techniques, such as power ultrasound use acoustic waves to induce physicochemical changes in the material. Some common applications of power ultrasound include its use to induce chemical reactions (sonochemistry) and to promote crystallization (sonocrystallization). These techniques have received the attention of several industries including pharmaceutical, chemical, and food industries given the advantages they offer. Ultrasound techniques are economically viable and relatively easy to incorporate into industrial operation. These techniques can be used to improve both reproducibility and yield of production; they are non-thermal and environmentally clean.

This Brief describes in detail recent research performed in the area of sonocrystallization of fats. Since this is a relatively new area of study, this Brief attempts to summarize research performed to date and provides a background of knowledge on which to build.

Some readers of this Brief may not be familiar with acoustic techniques and therefore an overview of ultrasound theory will be presented in Chaps. 2 and 3. A summary of the use of power ultrasound in the food industry is included in Chap. 4 and a description of the various physical phenomena involved in the use of power ultrasound to induce crystallization is presented in Chap. 5. A description of recent research in fat sonocrystallization and detailed information about experimental conditions used, such as type of equipment and ultrasound settings, will be presented in Chap. 6. The Brief will end with a discussion of the future trends in sonocrystallization of fats.

This Brief offers information to food producers and researchers interested in using power ultrasound as an additional processing tool to modify the physical characteristics of lipid-based foods. Graduate and undergraduate students interested in studying food processing may also benefit from this Brief.

References

Arends BJ, Blindt RA, Janssen J, Patrick M (2003) Crystallization process using ultrasound. US Patent 6,630,185 B2

Baxter JF, Morris GJ, Gaim-Marsoner G (1997a) Process for accelerating the polymorphic transformation of edible fats using ultrasonication. EP 0 765 605 A1

Baxter JF, Morris GJ, Gaim-Marsoner G (1997b) Process for retarding fat bloom in fat-based confectionery masses. EP 0 765 606 A1

Bonanome A, Grundy SM (1988) Effect of dietary stearic acid on plasma cholesterol and lipoprotein levels. N Engl J Med 318:1244–1248

Grande F, Andreson JT, Keys A (1970) Comparison of effects of palmitic and stearic acids in the diet on serum cholesterol in man. Am J Clin Nutr 23:1184–1193

Higaki K, Ueno S, Koyano K, Sato K (2001) Effects of ultrasonic irradiation on crystallization behavior of tripalmitoylglycerol and cocoa butter. J Am Oil Chem Soc 78:513–518

Martini S, Awad T, Marangoni AG (2006) Structure and properties of fat crystal networks. In: Gunstrone FD (ed) Modifying lipids for use in food. CRC Press, New York, pp 142–169

Martini S, Suzuki AH, Hartel RW (2008) Effect of high intensity ultrasound on crystallization behavior of anhydrous milk fat. J Am Oil Chem Soc 85:621–628

Martini S, Tejeda-Pichardo R, Ye Y, Padilla SG, Shen FK, Doyle T (2012) Bubble and crystal formation in lipid systems during high-intensity insonation. J Am Oil Chem Soc 89:1921–1928. doi:10.1007/s11746-012-2085-z

Mensink RP, Plat J (2008) Dietary fats and coronary heart disease. In: Akoh CC and Min DB (eds) Food lipids: chemistry, nutrition, and biotechnology. CRC Press, New York, pp 551–577

Mensink RP, Zock PL, Kester ADM, Katan MB (2003) Effects of dietary fatty acids and carbohydrates on the ratio of serum total to HDL cholesterol and on serum lipid and apolipoproteins: a meta-analysis of 60 controlled trials. Am J Clin Nutr 77:1146–1155

O'Brien RD (2000) Shortening technology. In: O'Brien RD, Farr WE, Wan PJ (eds) Introduction to fats and oil technology, 2nd edn. AOCS Press, Champaign, IL, pp 421–451

O'Brien RD (2009) Shortening types. In: O'Brien RD (ed) Fats and oils: formulating and processing for applications. CRC Press, New York, pp 347–360

Patrick M, Blindt R, Janssen J (2004) The effect of ultrasonic intensity on the crystal structure of palm oil. Ultrason Sonochem 11:251–255

Peyronel MF, Acevedo NC, Marangoni AG (2010) Structural and mechanical properties of fats and their implications for food quality. In: Skibsted LH, Risbo J, Andersen ML (eds) Chemical deterioration and physical instability of food and beverages. CRC Press, New York, pp 216–259

Suzuki A, Lee J, Padilla S, Martini S (2010) Altering functional properties of fats using power ultrasound. J Food Sci 75:E208–E214

Ueno S, Ristic RI, Higaki K, Sato K (2003) In situ studies of ultrasound-simulated fat crystallization using synchrotron radiation. J Phys Chem B 107:4927–4935

Ye Y, Wagh A, Martini S (2011) Using high intensity ultrasound as a tool to change the functional properties of interesterified soybean oil. J Agric Food Chem 59:10712

Chapter 2
An Overview of Ultrasound

2.1 Wave Propagation

2.1.1 General Aspects

Sound is a waveform that propagates away from a source in an elastic media generating density variations. The generation and perception of sound is a consequence of the transmission of mechanical energy through the media. That is, a device that generates sound must perform some kind of mechanical work to a medium. In the case of human voice, sound is generated by the mechanical work produced by the vocal cords in the air (medium of propagation). Similarly, to detect sound, mechanical work must be applied to the detector. In the human hearing system mechanical work is achieved through vibrations in the eardrum. Since mechanical work or energy is associated with the transmission of sound, the elastic and inertial properties of the material in which sound propagation occurs will affect the efficiency of wave propagation. For example, sound propagation in air is slower than in water with values of 343 and 1,497 m s^{-1}, respectively (Leighton 1994).

Acoustic waves are characterized by their frequency (cycles per second), wavelength (distance between cycles), and amplitude (height of the wave). Depending on the frequency of the waveform, sound waves can be classified as *infrasonic, sonic,* and *ultrasonic*. Sonic waves have frequencies between 20 and 20,000 Hz, which correspond to the frequency range of the human hearing. Waves with frequencies below 20 Hz are classified as *infrasonic*, while waves with frequencies above 20,000 Hz are classified as ultrasonic. The detection of sound in nature occurs over the entire spectrum of frequencies where certain animals can detect acoustic waves in the infrasonic range; while others detect sounds in the ultrasonic range. The frequency range of sound detection of different species, including humans, is detailed in Fig. 2.1a. The maximum and minimum frequency values of the detection range are reported in this figure. For example, elephants can detect sounds in the range of 5–12,000 Hz, while moths can detect sounds in a significantly smaller range that include frequencies between 20,000 and 50,000 Hz. Bats and porpoises can detect sounds with frequencies as high as

S. Martini, *Sonocrystallization of Fats*, SpringerBriefs in Food, Health, and Nutrition,
DOI: 10.1007/978-1-4614-7693-1_2, © Silvana Martini 2013

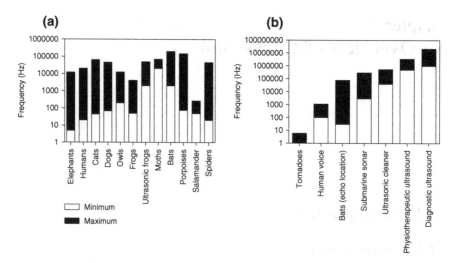

Fig. 2.1 Detection (**a**) and emission (**b**) acoustic frequencies of common acoustic equipment, and different species and phenomena found in nature. The *white* section of the *bar* represents the minimum frequency and the *black* section of the *bar* represents the maximum frequency of detection and emission. Reprinted with modifications from Leighton (1994), with permission from Elsevier

200,000 and 150,000 Hz, respectively, while salamanders can detect sounds at a maximum frequency of 220 Hz (Leighton 1994). Animals not only detect sound, they also emit sound over a wide range of frequencies (Fig. 2.1b). Bats emit sounds of frequencies in the range of 30–80,000 Hz to locate and identify objects. On the other extreme of the frequency scale, acoustic waves, with frequencies between 1 and 10 MHz are commonly used in diagnostic ultrasound techniques (Leighton 1994).

2.1.2 Wave Propagation

When acoustic waves travel through a material they do so at a specific velocity. This velocity is determined by the frequency and wavelength of the wave. Equation (2.1) describes the relationship between acoustic velocity, frequency, and wavelength.

$$c = v\lambda \tag{2.1}$$

where c is the acoustic velocity [m s^{-1}], v is the acoustic frequency [s^{-1}], and λ is the wavelength [m] of the acoustic wave. As previously mentioned, the speed of sound is affected by the characteristics of the material through which the sound is being propagated. If the sound propagates in a liquid or a gas, the speed of sound is a function of the bulk modulus of the material (Eq. 2.2):

$$c = \sqrt{\frac{K}{\rho}} \qquad (2.2)$$

where K is the bulk modulus and ρ is the density. When sound waves propagate in the solid, the Young (or elastic) modulus is used instead. The relationship between the Young (E) and bulk modulus (K) is shown in Eq. (2.3) (McClements 1991).

$$E = K + \frac{4}{3}G \qquad (2.3)$$

where G is the shear modulus. The change in acoustic velocity as a function of the material properties has been used to evaluate the chemical compositions of vegetable oils (McClements and Povey 1988a, b, 1992), to quantify the structural and mechanical properties of fats (Maleky et al. 2007), to evaluate the rheological behavior of xanthan/sucrose mixtures (Saggin and Coupland 2004a, b), and to evaluate ice formation in frozen food systems (Gülseren and Coupland 2007b).

Table 2.1 shows the speed of sound values in different materials. In general, the speed of sound is the lowest in gases with values in the range of 200–500 m s^{-1}, followed by the speed of sound in liquids, with values in the range of 1,200–2,000 m s^{-1}. The highest values of speed of sound are found in solids, with values in the range of 3,200–6,500 m s^{-1}.

Acoustic waves propagate in the media through *longitudinal or transversal waves* that generate localized displacement of particles or molecules present in the material. These two types of waves differ in the direction of particle displacement that occurs during wave propagation. In longitudinal waves particles are displaced parallel to the direction of the wave, while in transversal waves particles are displaced perpendicular to the direction of the wave. Acoustic waves are usually longitudinal but they can be transversal when they propagate through solids. It is important to note here that when acoustic waves travel through a material they generate only local displacement of particles; there is no movement of particles from one point to the other. Instead, particles oscillate around their equilibrium position as a consequence of wave propagation from the source to the detector. To better understand this concept, Fig. 2.2 shows a typical acoustic wave and the relationship between acoustic pressure and particle displacement (Leighton 1994).

Table 2.1 Speed of sound–c–(m s^{-1}) of some common materials

Solids		Liquids		Gases	
Material	c (m s^{-1})	Material	c (m s^{-1})	Material	c (m s^{-1})
Aluminum[a]	6,400	Ethyl alcohol	1,207	Air (0 °C)	331
Cork	500	Distilled water	1,497	Air (20 °C)	343
Pyrex glass	5,640	Sea water	1,531	Carbon dioxide	259
Gold	3,240	Mercury	1,450	Water vapor (134 °C)	494
Maple wood	4,110	Glycerol	1,904	Nitrogen	334
Stainless steel	5,790	Castor oil[a]	1,500	Oxygen	316

[a] Leighton (1994)

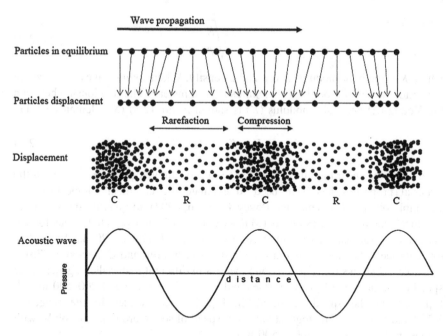

Fig. 2.2 Diagram of the propagation of longitudinal acoustic waves. Wave propagation occurs from *left* to *right* generating the displacement of particles around their equilibrium position. This displacement of particles results in zones of high (compression) and low (rarefaction) density zones in the material that correspond to high and low amplitudes in the acoustic wave, respectively. Reprinted with modifications from Leighton (1994), with permission from Elsevier

The first line of Fig. 2.2 shows particles in equilibrium position. When an acoustic wave travels from left to right in the direction shown by the arrow, particles move around their equilibrium position generating zones of compression and rarefaction. As already mentioned, no net displacement of particles is observed. The oscillation of particles results in gradients in the media where zones of high and low concentration of particles are observed. Highly concentrated zones of particles are observed when the media is compressed, while a low concentration of particles is observed in the rarefaction zones. The compression and rarefaction zones correspond to maximum and minimum amplitudes in the acoustic wave, respectively, as shown in the third line of Fig. 2.2 (Leighton 1994).

2.1.3 Introduction to Ultrasonic Techniques

Within the ultrasound range, acoustic techniques can be categorized into different groups according to the frequency and power of operation. High frequency and low power techniques are classified into *diagnostic* and *high frequency ultrasound*, while low frequency and high power techniques are classified into *power ultrasound*. Note that the category of diagnostic ultrasound includes techniques that use higher frequencies than those used by techniques included in the category of high frequency ultrasound. These techniques will be described in the following sections.

2.2 Diagnostic Ultrasound

Diagnostic ultrasound includes a series of pulse-echo techniques commonly used by the medical industry to evaluate the state of internal tissue structures (medical imaging). These are non-invasive, low power (<100 mW cm^{-2}), high frequency (1–10 MHz) techniques. Acoustic waves used for diagnostic applications are so low in intensity or power that they do not impart change to the physicochemical properties of the tissue.

Ultrasound is also used in the medical industry as a therapeutic tool. Low intensity (~ 1 W cm^{-2}) ultrasound is used as a deep-heating agent and ultrasound with higher intensities (~ 10 W cm^{-2}) can be used to treat oncological diseases. Finally, significantly higher power intensities (10^3 W cm^{-2}) are used in short duration pulses to modify body tissues (Frizzell 1988).

2.3 High Frequency Ultrasound

High frequency ultrasound techniques use frequencies between 100 kHz and 1 MHz, which are lower than those used in diagnostic ultrasound. High frequency ultrasound also use low power levels and therefore they do not induce changes in the material. High frequency ultrasound has been used extensively in several food science applications including monitoring of crystallization of lipids (McClements and Povey 1987, 1988a; Singh et al. 2004; Saggin and Coupland 2002; Santacatalina et al. 2011; Martini et al. 2005a, b, c), characterizing edible oils and fats (McClements and Povey 1988b, 1992), predicting viscoelastic properties of the material (Saggin and Coupland 2001; 2004a, b; Maleky et al. 2007; Mert and Campanella 2007), characterizing emulsions and suspensions (McClements 1991; McClements et al. 1990; McClements and Povey 1989; Coupland and McClements 2001), monitoring crystallization of lipids in emulsions (Hodate et al. 1997; Kashchiev et al. 1998; Kaneko et al. 1999; Vanapalli and Coupland 2001; Gülseren and Coupland 2007a, b; McClements et al. 1993), monitoring dissolution and crystallization of carbohydrates (Gülseren and Coupland 2008; Yucel and Coupland 2010, 2011a, b), and monitoring gel formation (Benguigui et al. 1994; Audebrand et al. 1995; Corredig et al. 2004; Dwyer et al. 2005; Wan et al. 2007) to name a few. A significant advantage of this technique over other tools used to characterize materials is that it is non-invasive, non-destructive, and can be used in concentrated and opaque materials.

2.4 Power Ultrasound

Power ultrasound or high intensity ultrasound uses significantly lower frequencies than those used in diagnostic and high frequency ultrasound. Frequencies used in power ultrasound range between 20 and 100 kHz. When low frequencies and high power ($10-10,000$ W cm^{-2}) acoustic waves such as those used in these techniques travel through a medium they induce formation of cavities (Rastogui 2011; Bermudez-Aguirre and Barbosa Canovas 2011). Acoustic cavitation and the events associated with this phenomenon are responsible for inducing several physico-chemical changes in the material. These events will be discussed in detail in Chap. 5.

Power ultrasound is a highly invasive technique that uses high intensity acoustic waves to purposely change the properties of materials. Some examples of the use of power ultrasound include therapeutic medicine such as physiotherapy, chemotherapy, ultrasonic thrombolysis, and drug delivery (Mason 2011). Other common uses of power ultrasound include inducing crystallization of materials (sonocrystallization) and inducing or causing chemical reactions (sonochemistry) that would not occur in the absence of ultrasound (Petrirt et al. 1994; Suslick et al. 1997; Gedanken 2004; Mason 1999; Kasaai et al. 2008; Wu et al. 2008).

Sonocrystallization has been used extensively by industries including pharmaceutical (Luque de Castro and Priego-Capote 2007; Krishna et al. 2007; Miyasaka et al. 2006; Louhi-Kultanen et al. 2006; Manish et al. 2005; McCausland 2007), chemical (Mansour and Takrouri 2007; Bucar and MacGillivray 2007; Paradkar et al. 2006; Li et al. 2003, 2006; Ruecroft et al. 2005; Oldenburg et al. 2005; Kaerger and Price 2004; Cains et al. 1998) and food (Bund and Pandit 2007a, b).

In addition to the use of power ultrasound in sonochemistry and sonocrystallization, this technique has recently been used to change the physicochemical properties of proteins (Villamiel and de Jong 2000; Kresic et al. 2008; Jambrak et al. 2008, 2009; Ashokkumar et al. 2009; Martini et al. 2010; Zisu et al. 2010, 2011; Arzeni et al. 2012; Martini and Walsh 2012; Hu et al. 2013). Power ultrasound has been used in food protein suspensions to increase the clarity of whey suspensions, to increase protein solubility, to change suspension viscosity, to decrease particle size, and to narrow particle size distributions. The source and concentration of sonicated protein determine the nature of change induced by sonication.

The use of ultrasonic techniques has caught the attention of the industry given the many advantages offered. Some benefits include: (a) improvements in product quality, (b) reduction in food processing costs, (c) decreased processing times, (d) reduced chemical and physical hazards, (e) environmentally friendly and sustainable, and (f) scalable throughput (Patist and Bates 2008; Chemat et al. 2011).

References

Arzeni C, Perez OE, Pilosof AMR (2012) Functionality of egg white proteins as affected by high intensity ultrasound. Food Hydrocoll 29:308–316

Ashokkumar M, Lee J, Zisu B, Bhaskarcharya R, Palmer M, Kenrish S (2009) Hot topic: sonication increases the heat stability of whey proteins. J Dairy Sci 92:5353–5356

Audebrand M, Doublier JL, Durand D, Emery JR (1995) Investigation of gelation phenomena of some polysaccharides by ultrasonic spectroscopy. Food Hydrocoll 9:195–203

Benguigui L, Emery J, Durand D, Busnel JP (1994) Ultrasonic study of milk-clotting. Lait 74:197–206

Bermudez-Aguirre D, Barbosa Canovas GV (2011) Chapter 16: power ultrasound to process dairy products. In: Feng H, Barbosa-Canovas GV, Weiss J (eds) Ultrasound technologies for food and bioprocessing. Springer, New York, pp 445–465

Bucar DK, MacGillivray LR (2007) Preparation and reactivity of nanocrystalline cocrystals formed via sonocrystallization. J Am Chem Soc 129:32–33

Bund RK, Pandit AB (2007a) Rapid lactose recovery from paneer whey using sonocrystallization: a process optimization. Chem Eng and Proc 46:846–850

Bund RK, Pandit AB (2007b) Sonocrystallization: effect on lactose recovery and crystal habit. Ultrason Sonochem 14:143–152

Cains PW, Martin PD, Price CJ (1998) The use of ultrasound in industrial chemical synthesis and crystallization. 1. Applications to synthetic chemistry. Org Process Res Dev 2:34–48

Chemat F, Huma Z, Khan MK (2011) Applications of ultrasound in food technology: processing, preservation and extraction. Ultrason Sonochem 18:813–835

Corredig M, Alexander M, Dalgleish DG (2004) The application of ultrasonic spectroscopy to the study of the gelation of milk components. Food Res Int 37:557–565

Coupland JN, McClements DJ (2001) Droplet size determination in food emulsions: comparison of ultrasonic and light scattering methods. J Food Eng 50:117–120

Dwyer C, Donnelly L, Buckin V (2005) Ultrasonic analysis of rennet induced pregelation and gelation processes in milk. J Dairy Res 72(303):310

Frizzell LA (1988) Biological effect of acoustic cavitation. In: Suslick KS (ed) Ultrasound: its chemical, physical, and biological effects. VCH Publishers Inc., Weinheim, pp 287–303

Gedanken A (2004) Using sonochemistry for the fabrication of nanomaterials. Ultrason Sonochem 11:47–55

Gülseren I, Coupland JN (2007a) The effect of emulsifier type and droplet size on phase transitions in emulsified even-numbered n-alkanes. J Am Oil Chem Soc 84:621–629

Gülseren I, Coupland JN (2007b) Ultrasonic velocity measurements in frozen model food solutions. J Food Eng 79:1071–1078

Gülseren I, Coupland JN (2008) Ultrasonic properties of partially frozen sucrose solutions. J Food Eng 89:330–335

Hodate Y, Ueno S, Yano J, Katsuragi T, Tezuka Y, Tagawa T, Yoshimoto N, Sato K (1997) Ultrasonic velocity measurement of crystallization rates of palm oil in oil-water emulsions. Coll Surf A 128:217–224

Hu H, Wu J, Li-Chan ECY, Zhu L, Zhang F, Xu X, Fan G, Wang L, Huang X, Pan S (2013) Effects of ultrasound on structural and physical properties of soy protein isolate (SPI) dispersions. Food Hydrocoll 30:647–655

Jambrak AR, Mason TJ, Lelas V, Herceg Z, Herceg IL (2008) Effect of ultrasound treatment on solubility and foaming properties of whey proteins suspensions. J Food Eng 86:281–287

Jambrak AR, Lelas V, Mason TJ, Kresic G, Badanjak M (2009) Physical properties of ultrasound treated soy proteins. J Food Eng 93:386–393

Kaerger JS, Price R (2004) Processing of spherical crystalline particles via a novel solution atomization and crystallization by sonication (SAXS) technique. Pharm Res 21:372–381

Kaneko N, Horie T, Ueno S, Yano J, Katsuragi T, Sato K (1999) Impurity effects on crystallization rates of n-hexadecane in oil-in-water emulsions. J Cryst Growth 197:263–270

Kasaai MR, Arul J, Charlet G (2008) Fragmenetation of chitosan by ultrasonic irradiation. Ultrason Sonochem 15:1001–1008

Kashchiev D, Kaneko N, Sato K (1998) Kinetics of crystallization in polydisperse emulsions. J Coll Interface Sci 208:167–177

Kresic G, Lelas V, Jamrak AR, Herceg Z, Brncic SR (2008) Influence of novel food processing technologies on the rheological and thermophysical properties of whey proteins. J Food Eng 87:64–73

Krishna MV, Babu JR, Latha PVM, Sankar DG (2007) Sonocrystallization: for better pharmaceutical crystals. Asian J Chem 19:1369–1374

Leighton TG (1994) Chapter 1: the sound field. In: Leighton TG (ed) The acoustic bubble. Academic Press, New York, pp 1–66

Li H, Wang JK, Bao Y, Guo ZC, Zhang MY (2003) Rapid sonocrystallization in the salting-out process. J Cryst Growth 247:192–198

Li H, Li HR, Guo ZC, Liu Y (2006) The application of power ultrasound to reaction crystallization. Ultrason Sonochem 13:359–363

Louhi-Kultanen M, Karjalainen M, Rantanen J, Huhtanen M, Kallas J (2006) Crystallization of glycine with ultrasound. Int J Pharm 320:23–29

Luque de Castro MD, Priego-Capote F (2007) Ultrasound-assisted crystallization (sonocrystallization). Ultrason Sonochem 14:717–724

Maleky F, Campos R, Marangoni AG (2007) Structural and mechanical properties of fats quantified by ultrasonics. J Am Oil Chem Soc 84:331–338

Manish M, Harshal J, Anant P (2005) Melt sonocrystallization of ibuprofen: effect on crystal properties. Eur J Pharm Sci 25:41–48

Mansour AR, Takrouri KJ (2007) A new technology for the crystallization of dead sea potassium chloride. Chem Eng Commun 194:803–810

Martini S, Walsh MK (2012) Sensory characteristics and functionality of sonicated whey. Food Res Int 49:694–701

Martini S, Bertoli C, Herrera ML, Neeson I, Marangoni AG (2005a) In-situ monitoring of solid fat content by means of p-NMR and ultrasonics. J Am Oil Chem Soc 82:305–312

Martini S, Herrera ML, Marangoni AG (2005b) New technologies to determine solid fat content on-line. J Am Oil Chem Soc 82:313–317

Martini S, Bertoli C, Herrera ML, Neeson I, Marangoni AG (2005c) Attenuation of ultrasonic waves: influence of microstructure and solid fat content. J Am Oil Chem Soc 82:319–328

Martini S, Potter R, Walsh MK (2010) Optimizing the use of high intensity ultrasound to decrease turbidity in whey protein suspensions. Food Res Int 43:2444–2451

Mason TJ (1999) Sonochemistry: current uses and future prospects in the chemical and processing industries. Philos T R Soc A 357:355–369

Mason TJ (2011) Therapeutic ultrasound and overview. Ultrason Sonochem 18:847–852

McCausland LJ (2007) Production of crystalline materials by using high intensity ultrasound. US 7,244,307 B2

McClements DJ (1991) Ultrasonic characterization of emulsions and suspensions. Adv Coll Interface 37:33–72

McClements DJ, Povey MJW (1987) Solid fat content determination using ultrasonic velocity measurements. Int J Food Sci Tech 22:491–499

McClements DJ, Povey MJW (1988a) Comparison of pulsed NMR and ultrasonic velocity techniques for determining solid fat contents. Int J Food Sci Tech 23:159–170

McClements DJ, Povey MJW (1988b) Ultrasonic velocity measurements in some liquid triglycerides and vegetable oils. J Am Oil Chem Soc 65:1787–1790

McClements DJ, Povey MJW (1989) Scattering of ultrasound by emulsions. J Phys D Appl Phys 22:38–47

McClements DJ, Povey MJW (1992) Ultrasonic analysis of edible fats and oils. Ultrasonics 30:383–388

McClements DJ, Povey MJW, Betsanis E (1990) Ultrasonic characterization of a food emulsion. Ultrasonics 28:266–272

McClements DJ, Povey MJW, Dickinson E (1993) Absorption and velocity dispersion due crystallization and melting of emulsion droplets. Ultrasonics 31:433–437

Mert B, Campanella OH (2007) Monitoring the rheological properties and solid content of selected food materials contained in cylindrical cans using audio frequency sound waves. J Food Eng 79:546–552

Miyasaka E, Ebihara S, Hirasawa I (2006) Investigation of primary nucleation phenomena of acetylsalicylic acid crystals induced by ultrasonic irradiation—ultrasonic energy needed to activate primary nucleation. J Cryst Growth 295:97–101

Oldenburg K, Pooler D, Scudder K, Lipinski C, Kelly M (2005) High throughput sonication: evaluation for compound solubilization. Comb Chem High T Scr 8:499–512

Paradkar A, Maheshwari M, Kamble R, Grimsey I, York P (2006) Design and evaluation of celecoxib porous particles using melt sonocrystallization. Pharm Res 23:1395–1400

Patist A, Bates D (2008) Ultrasonic innovations in the food industry: from the laboratory to commercial production. Innovat Food Sci Emerg Technol 9:147–154

Petrirt C, Lamy M, Francony A, Benahcene A, David B (1994) Sonochemical degradation of phenol in dilute aqueous solutions: comparison of the reaction rates of 20 and 487 kHz. J Phys Chem 98:10514–10520

Rastogui NK (2011) Opportunity and challenges in application of ultrasound in food processing. Crit Rev Food Sci Nutr 51:705–722

Ruecroft G, Hipkiss D, Ly T, Maxted N, Cains PW (2005) Sonocrystallization: the use of ultrasound for improved industrial crystallization. Org Proc Res Dev 9:923–993

Saggin R, Coupland JN (2001) Oil viscosity measurement by ultrasonic reflectance. J Am Oil Chem Soc 78:509–511

Saggin R, Coupland JN (2002) Measurement of solid fat content by ultrasonic reflectance in model systems and chocolate. Food Res Int 35:999–1005

Saggin R, Coupland JN (2004a) Rheology of xanthan/sucrose mixtures at ultrasonic frequencies. J Food Eng 65:49–53

Saggin R, Coupland JN (2004b) Shear and longitudinal ultrasonic measurements of solid fat dispersions. J Am Oil Chem Soc 81:27–32

Santacatalina JV, Garice-Perez JV, Corona E, Benedito J (2011) Ultrasonic monitoring of lard crystallization during storage. Food Res Int 44:146–155

Singh AP, McClements DJ, Marangoni AG (2004) Solid fat content determination by ultrasonic velocimetry. Food Res Int 37:545–555

Suslick KS, Mdleleni MM, Ries JT (1997) Chemistry induced by hydrodynamic cavitation. J Am Chem Soc 119:9303–9304

Vanapalli SA, Coupland JN (2001) Emulsions under shear—the formation and properties of partially coalesced lipid structures. Food Hydrocoll 15:507–512

Villamiel M, de Jong P (2000) Influence of high-intensity ultrasound and heat treatment in continuous flow on fat, proteins and active enzymes in milk. J Agric Food Chem 48:472–478

Wan Q, Bulca S, Kulozik U (2007) A comparison of low-intensity ultrasound and oscillating rheology to assess the renneting properties of casein solutions after UHT heat pre-treatment. Int Dairy J 17:50–58

Wu T, Zivanovic S, Hayes DG, Weiss J (2008) Efficient reduction of chitosan molecular weight by high-intensity ultrasound: underlying mechanism and effect of process parameters. J Agric Food Chem 56:5112–5119

Yucel U, Coupland JN (2010) Ultrasonic characterization of lactose dissolution. J Food Eng 98:28–33

Yucel U, Coupland JN (2011a) Ultrasonic attenuation measurements of the mixing, agglomeration, and sedimentation of sucrose crystals suspended in oil. J Am Oil Chem Soc 88:33–38

Yucel U, Coupland JN (2011b) Ultrasonic characterization of lactose crystallization in gelatin gels. J Food Sci 76:E48–E54

Zisu B, Bhaskaracharya R, Kentish S, Ashokkumar M (2010) Ultrasonic processing of diary systems in large scale reactors. Ultrason Sonochem 17:1075–1081

Zisu B, Lee J, Chandrapala J, Bhaskaracharya R, Palmer M, Kentish S, Ashokkumar M (2011) Effect of ultrasound on the physical and functional properties of reconstituted whey protein powders. J Dairy Res 78:226–232

Chapter 3
Ultrasound Process Parameters

3.1 Acoustic Power, Intensity, and Attenuation

As mentioned, acoustic waves are characterized by their wavelength, velocity, and frequency. In addition to these intrinsic wave properties, different acoustic power levels can be used for specific applications. When a non-invasive technique is needed, low power levels are chosen, while higher powers are used when physicochemical changes in the material are required. It is important to define the terms acoustic power, intensity, and attenuation of an acoustic wave and understand the variables that affect these parameters.

Acoustic power is defined as the amount of acoustic energy (expressed in Joules) delivered per unit time (expressed in seconds), and it is therefore expressed in Watts. The strength of the acoustic wave can be expressed in terms of its *intensity*. In this case, the acoustic intensity is defined as the "wave energy rate crossing a unit area perpendicular to the direction of propagation" (Leighton 1994). Acoustic intensity is expressed in W cm^{-2} and can be described by the following equation:

$$I = \frac{P_A^2}{2Z} \tag{3.1}$$

and

$$Z = \rho c \tag{3.2}$$

where P_A is the acoustic wave pressure amplitude (Pa), Z is the impedance (kg m^{-2} s^{-1}), ρ is the density of the material (kg m^{-3}), and c is the acoustic velocity in the media (m s^{-1}). Equation 3.1 shows that the energy of acoustic waves is proportional to the acoustic pressure amplitude of the wave. That is, the higher the pressure amplitude of the wave, the higher the acoustic intensity generated. Therefore, some of the published work expresses the intensity of the wave by reporting pressure amplitudes rather than intensities.

In addition to using terms such as acoustic power and acoustic intensity, the term *intensity level* is sometimes used to characterize an acoustic wave. Intensity levels (IL) refer to the ratio of acoustic intensities of two different waves.

In general, IL are used to compare the intensity of an acoustic wave to the intensity of a reference wave using a logarithmic function (Eq. 3.3) and is expressed in decibel units (dB):

$$IL = 10 \log_{10} \left(\frac{I}{I_{ref}} \right)$$ (3.3)

where I is the intensity of the wave and I_{ref} is the intensity of a reference wave (Leighton 1994).

When an acoustic wave travels through a material some of the acoustic energy is lost as the wave propagates through the media. This loss of energy is called *attenuation* and can be expressed in terms of acoustic pressure using the following equation:

$$P = P_0 e^{i(\omega t - qx)} e^{-bx}$$ (3.4)

where P is the acoustic pressure of the wave at any position away from the source, P_0 is the acoustic pressure at the source, ω is the circular frequency of the wave ($2\pi v$), q is the wave number, i is $\sqrt{-1}$, t is time, b is the *amplitude attenuation constant or coefficient*, and x is the direction of propagation of the wave. Combining Eqs. (3.1) and (3.4), it is obvious that the intensity of the acoustic wave decays as a function of e^{-2bx}. Therefore, to understand the propagation of acoustic waves it is important to know the value of b (dB m^{-1}) in order to quantify the loss of acoustic energy during propagation in a specific material. Table 3.1 summarizes values of attenuation coefficients for several food materials ranging from a couple of dB m^{-1} to hundreds of dB m^{-1}. For example, the attenuation coefficient for water measured at 5 MHz is 5 dB m^{-1}, while the attenuation coefficient for 2 % Xanthan gel measured at 5 MHz is 170 dB m^{-1}. In addition, attenuation coefficients depend on the frequency of measurement because acoustic waves are more attenuated at higher frequencies.

The loss of energy (attenuation) experienced by an acoustic wave that propagates in a media is a result of various phenomena. The first phenomenon is the transformation of mechanical energy into heat due to work that the acoustic wave performs as it travels through a viscous medium. The second phenomenon responsible for attenuation is the dissipation of energy as a result of compression and rarefaction events during wave propagation. Last but not least, the scattering from inhomogeneities in the media is a significant cause of attenuation in a non-homogeneous medium. This phenomenon is especially important in food systems where particles, crystals, and/or droplets can be present in the media (Leighton 1994; Meyer et al. 2006).

The fact that ultrasonic waves are attenuated by inhomogeneities has been exploited by several food scientists who use low intensity ultrasound as an analytical tool to monitor physical changes in opaque materials. These physical changes include monitoring droplet sizes and distribution in emulsions (McClements 1991; McClements et al. 1990, 1993; McClements and Povey 1989; Kaneko et al. 1999; Hodate et al. 1997; Gülseren and Coupland 2007, 2008),

Table 3.1 Attenuation coefficients of different food materials expressed in dB m^{-1}. Since attenuation is a function of frequency, the frequency used to measure the attenuation coefficient is presented in parenthesis. Reprinted from McClements and Povey (1992), with permission from Elsevier. Reprinted from Povey (1989), with permission from Elsevier

Material	Attenuation coefficient (dB m^{-1})	Reference
Water[b]	5.8 (5 MHz)	Kaye and Laby (1986)
Proving bread dough[b]	30 (0.5 MHz)	Moorjani (1984)
Egg white[b]	45 (2.5 MHz)	Povey and Wilkinson 1980
2 % Xanthan gel[b]	170 (5 MHz)	Rahalkar et al. (1986)
40 % Gum arabic gel[b]	270 (10 MHz)	Pryor et al. (1958)
Homogenized milk[b]	400 (1 MHz)	Hueter et al. (1953)
Skimmed milk[b]	200 (1 MHz)	Hueter et al. (1953); Saraf et al. (1982)
White fish flesh[b]	61 (1 MHz)	Freese and Makow (1968)
Castor oil[a]	32.7 (2 MHz)	Gladwell et al. (1985)
Olive oil[a]	6.5 (2 MHz)	Gladwell et al. (1985)
Peanut oil[a]	6.0 (2 MHz)	Kuo (1971)
Safflower oil[a]	4.0 (2 MHz)	Gladwell et al. (1985)
Soybean oil[a]	5.5 (2 MHz)	Kuo (1971)
20 % lactose in water	5 (2.25 MHz)	Yucel and Coupland (2010), (2011a, b)
40 % sucrose solution in water (partially frozen)	100 (2.25 MHz)	Gülseren and Coupland (2008)

[a] Reprinted from McClements and Povey (1992), with permission from Elsevier
[b] Reprinted from Povey (1989), with permission from Elsevier

monitoring crystallization behavior in lipids (McClements and Povey 1987; Martini et al. 2005a, b, c; Singh et al. 2004; Santacatalina et al. 2011; Saggin and Coupland 2002, 2004) and sugars (Yucel and Coupland 2010, 2011a, b), and monitoring gelation of polysaccharides (Audebrand et al. 1995) and milk components (Corredig et al. 2004; Dwyer et al. 2005). Some of these researches also demonstrated higher attenuation levels at higher acoustic frequencies.

3.2 Calculation of Acoustic Power and Acoustic Intensity

3.2.1 Acoustic Power

The effects of ultrasound waves in different materials depend on the power level used. It is therefore important to report acoustic power levels used in an experiment. This is especially true for experiments that use power ultrasound or high intensity ultrasonic techniques. The power level generated by acoustic waves is inversely proportional to the square of the frequency (Power $\propto 1/f^2$) as described by Bermudez-Aguirre et al. (2011). This means that if high power levels are required, low frequencies must be used. Bench-top power ultrasound generators commonly used by food scientists are usually designed for cell disruption experiments and/or for emulsification purposes and operate at frequencies between 20 and 60 kHz. This type of equipment usually has a display panel that shows the electrical power delivered during the sonication experiment. Electrical power levels are used to indicate the energy delivered by the acoustic generator to the tip or transducer delivering acoustic signals to the sample. The different types of transducers and the difference between electrical and acoustic power will be discussed in Section 3.3.

Acoustic power levels obtained during sonication of a material can be calculated using Eq. (3.5) as described by Jambrak et al. (2008, 2009):

$$P = m \times C_p \times \frac{dT}{dt} \tag{3.5}$$

where P is the acoustic power in W, m the mass of the sonicated sample expressed in g, C_p the specific heat capacity of the medium at constant pressure, expressed in $J\ g^{-1}\ °C^{-1}$, and (dT/dt) is the temperature increase during sonication expressed in $°C\ s^{-1}$. The C_p of the sample can be calculated using differential scanning calorimetry using sapphire as the reference (known C_p).

Martini et al. (2010, 2012) reported acoustic power levels during the sonication of soybean oil (SBO), interesterified SBO (Ye et al. 2011) and whey solutions. A detailed analysis of temperature changes during sonication was reported by Martini et al. (2012) and the results are summarized in Fig. 3.1. This figure shows temperature profiles of SBO sonicated for different periods of time (5, 10, and 60 s) at different electrical powers (6, 21, 42, 66, and 90 W). Several observations can be made from this figure. First, a rapid increase in temperature was observed in the samples immediately after sonication was applied. This increase in temperature was linear when low powers were used and reached a plateau when higher power levels were used. Similar temperature profiles were reported by Ter Haar (1988) who suggests that the temperature plateau is reached when there is a balance between the energy delivered to the system and the energy dissipated by the media. The second significant observation from Fig. 3.1 is that the increase in temperature (ΔT) in the sonicated sample is more significant when ultrasound is applied for longer periods of time. When SBO samples are sonicated for 5, 10, and 60 s, ΔT values range between 0 and 3 °C, 0 and 4 °C, and 1 and 16 °C, respectively. Considering the C_p value of SBO ($2.05 \pm 0.02\ J\ g^{-1}\ °C^{-1}$) and using Eq. (3.5), these temperature increases correspond to acoustic power levels between 1 and 70 W. It is important to note that the acoustic power achieved during sonication also depends on the duration of the signal. In addition, data reported by Martini et al. (2012) show that lower electrical power does not always result in a lower acoustic power. When low electrical powers are used, the acoustic energy is concentrated closer to the tip resulting in localized temperature increases and therefore high acoustic power is recorded. When higher electrical power levels are

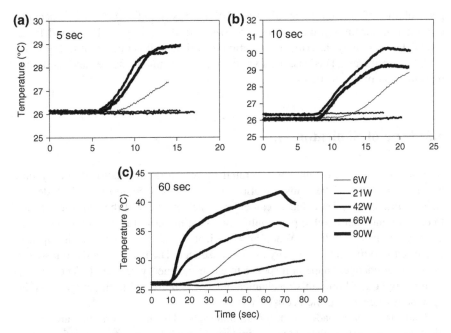

Fig. 3.1 Effect of high intensity ultrasound application on temperature increases in soybean oil (SBO). SBO was sonicated for 5 (**a**), 10 (**b**), or 60 (**c**) sec at 26 °C. Reprinted from Martini et al. (2012), with kind permission from Springer Science and Business Media

used, the acoustic energy dissipates through the media resulting in an overall lower increase in temperature and therefore a lower acoustic power.

3.2.2 Acoustic Intensity

In addition to reporting acoustic power levels it is important to report acoustic intensities and acoustic densities so that it is easier to compare studies. As mentioned, the acoustic intensity is the acoustic power generated by the tip per unit area and it is expressed in W cm^{-2}. The calculation of the acoustic intensity will therefore depend on the geometry of the transducer used. If the transducer has a circular area, then πr^2 will be used to calculate the power intensity (Bermudez-Aguirre et al. 2011), where r is the radius of the tip. Acoustic intensity increases as the tip size decreases. This means that bigger tips will result in lower power intensities than smaller and thinner tips.

Other units of measurement commonly used to describe the intensity of acoustic waves delivered to the material include *power density* and *energy input*. Power density measures the acoustic power delivered as a function of the volume of the sonicated material (W cm^{-3}). Energy input measures the total energy input per volume of treated material (kWh/L). Energy input values are especially important in scaling-up ultrasound processes into the industrial (Patist and Bates 2011). To accurately describe a sonication experiment, it is important to report all of these parameters. This ensures a valid comparison among studies and facilitates reproduction of experiments.

3.3 Type of Sonicator Tip

Acoustic waves in a media are generated by transforming electrical energy into mechanical energy. The transformation of energy is accomplished by a device called a *transducer*. Transducers can either generate or detect the acoustic signals. In experiments that involve power ultrasound techniques, it is common to refer to these transducers as "tips". When an electrical signal is delivered to the tip it is converted into mechanical energy in the transducer. The material in the transducer oscillates or changes shape in response to electrical energy delivered. When the tip is submerged in a liquid this mechanical energy is delivered to the medium as the tip vibrates generating the acoustic wave.

Transducers are made of different materials. The most common are *piezo-electric* transducers. Piezoelectric materials, such as quartz, lithium sulfate zirconate titanates, are characterized by electric charges fixed in the crystal lattice. When pressure is applied to these crystals, equal and opposite electric charges are generated on opposite faces of the crystal resulting in an electrical potential between them creating a displacement of crystal planes. The production of electric charges in response to mechanical stress is reversible. Therefore, when an electric voltage is applied to the material its shape will be altered transforming the electric energy into mechanical energy. If the electrical signal is applied in an oscillatory manner, for example in a sinusoidal way, the material will change its shape following the same frequency as the electric signal. When this material is coupled to a liquid, the geometrical oscillation of the transducer generates an acoustic wave in the liquid that maintains the same frequency as the electrical signal (Shoh 1988; Leighton 1994). The amplitude of mechanical deformation or vibration depends on the energy delivered to the tip; the higher the energy, the higher the amplitude of vibration and the higher the acoustic power generated. It is therefore very common to use the amplitude of vibration of the tip as a measurement of the acoustic power delivered to the media. Tip amplitude vibration values are provided by the manufacturer of the ultrasound equipment and depend on the type of tip and on the amount of energy delivered to the tip.

It is necessary to emphasize the difference between electrical and acoustic power. Commonly used sonicators display the electrical power delivered to the tip to achieve a specific amplitude of vibration at the tip. However, these electrical power levels do not correspond to the actual acoustic power in the media that is being sonicated. The electrical power needed to achieve a specific amplitude will vary depending on the type of tip used and on the material that is being sonicated. For example, if a viscous material is being sonicated, a higher electrical power is needed to achieve a specific amplitude of vibration at the tip. Therefore, electrical power levels are not very useful to characterize sonication conditions, and it is more useful to report actual acoustic power levels.

The second type of transducer operates under a mechanism called *magnetostriction*. This mechanism is based on a length change in a ferromagnetic material caused by an applied magnetic field. The magnetic field can oscillate in a sinusoidal manner generating a length change in the material with the same oscillation characteristics as the applied magnetic field. Magnetostrictive transducers are efficient only at low frequencies (<30 kHz) (Shoh 1988; Leighton 1994) and are very brittle and therefore have been gradually replaced by piezoelectric transducers.

In summary, several process parameters must be considered when performing experiments using power ultrasound techniques. It is important to understand and correctly report sonication parameters such as acoustic frequency, power, intensity, and density together with sample volume, tip type, and dimensions. Variations in any of these parameters will affect the results obtained.

References

Audebrand M, Doublier JL, Durand D, Emery JR (1995) Investigation of gelation phenomena of some polysaccharides by ultrasonic spectroscopy. Food Hydrocolloid 9:195–203

Bermudez-Aguirre D, Mobbs T, Barbosa-Canovas G (2011) Ultrasound applications in food processing. In: Feng H, Barbosa-Canovas GV, Weiss JW (eds) Ultrasound technologies for food and bioprocessing. Springer, New York, pp 65–105

Corredig M, Alexander M, Dalgleish DG (2004) The application of ultrasonic spectroscopy to the study of the gelation of milk components. Food Res Int 37:557–565

Dwyer C, Donnelly L, Buckin V (2005) Ultrasonic analysis of rennet induced pregelation and gelation processes in milk. J Dairy Res 72:303–310

Freese M, Makow D (1968) High-frequency ultrasonic properties of fresh water fish tissue. J Acoust Soc Am 44:1282–1289

Gladwell N, Javanaud C, Peers KE, Rahalkar RR (1985) Ultrasonic behavior of edible oils: correlation with rheology. J Am Oil Chem Soc 62:1231–1236

Gülseren I, Coupland JN (2007) The effect of emulsifier type and droplet size on phase transitions in emulsified even-numbered n-alkanes. J Am Oil Chem Soc 84:621–629

Gülseren I, Coupland JN (2008) Ultrasonic properties of partially frozen sucrose solutions. J Food Eng 89:330–335

Hodate Y, Ueno S, Yano J, Katsuragi T, Tezuka Y, Tagawa T, Yoshimoto N, Sato K (1997) Ultrasonic velocity measurement of crystallization rates of palm oil in oil-water emulsions. Colloid Surf A 128:217–224

Hueter TF, Morgan H, Cohen MS (1953) Ultrasonic attenuation in biological suspensions. J Acoust Soc Am 25:1200–1201

Jambrak AR, Mason TJ, Lelas V, Herceg Z, Herceg IL (2008) Effect of ultrasound treatment on solubility and foaming properties of whey protein suspensions. J Food Eng 86:281–287

Jambrak AR, Lelas V, Mason TJ, Krešic G, Badanjak M (2009) Physical properties of ultrasound treated soy proteins. J Food Eng 93:386–393

Kaneko N, Horie T, Ueno S, Yano J, Katsuragi T, Sato K (1999) Impurity effects on crystallization rates of n-hexadecane in oil-in-water emulsions. J Cryst Growth 197:263–270

Kaye GWC, Laby TH (1986) In: Kaye GWC, Laby TH (eds) Tables of physical and chemical constants, 15th edn. Longman, London, p 30

Kuo HL (1971) Variation of ultrasonic velocity and absorption with temperature and frequency in high viscosity vegetable oils. Jpn J Appl Phys 10:167–170

Leighton TG (1994) The sound field. In: Leighton TG (ed) The acoustic bubble. Academic Press, New York, pp 1–66

Martini S, Bertoli C, Herrera ML, Neeson I, Marangoni AG (2005a) In-situ monitoring of solid fat content by means of p-NMR and ultrasonics. J Am Oil Chem Soc 82:305–312

Martini S, Herrera ML, Marangoni AG (2005b) New technologies to determine solid fat content on-line. J Am Oil Chem Soc 82:313–317

Martini S, Bertoli C, Herrera ML, Neeson I, Marangoni AG (2005c) Attenuation of ultrasonic waves: Influence of microstructure and solid fat content. J Am Oil Chem Soc 82:319–328

Martini S, Tejeda-Pichardo R, Ye Y, Padilla SG, Shen FK, Doyle T (2012) Bubble and crystal formation in lipid systems during high-intensity insonation. J Am Oil Chem Soc 89:1921–1928

Martini S, Potter R, Walsh MK (2010) Optimizing the use of high intensity ultrasound to decrease turbidity in whey protein suspensions. Food Res Int 43:2444–2451

McClements DJ, Povey MJW (1987) Solid fat content determination using ultrasonic velocity measurements. Int J Food Sci Tech 22:491–499

McClements DJ, Povey MJW (1989) Scattering of ultrasound by emulsions. J Phys D Appl Phys 22:38–47

McClements DJ (1991) Ultrasonic characterization of emulsions and suspensions. Adv Colloid Interfac 37:33–72

McClements DJ, Povey MJW (1992) Ultrasonic analysis of edible fats. Ultrasonics 30:383–388

McClements DJ, Povey MJW, Betsanis E (1990) Ultrasonic characterization of a food emulsion. Ultrasonics 28:266–272

McClements DJ, Povey MJW, Dickinson E (1993) Absorption and velocity dispersion due crystallization and melting of emulsion droplets. Ultrasonics 31:433–437

Meyer S, Rajendram VS, Povey MJW (2006) Characterization of reconstituted milk powder by ultrasound spectroscopy. J Food Quality 29:405–418

Moorjani R (1984) An investigation into the acoustics of bread doughs. MSc thesis, University of Leeds.

Patist A, Bates D (2011) Industrial applications of high power ultrasonics. In: Feng H, Barbosa-Canovas GV, Weiss JW (eds) Ultrasound technologies for food and bioprocessing. Springer, New York, pp 599–616

Povey MJW (1989) Ultrasonics in food engineering. J Food Eng 9:1–20

Povey MJW, Wilkinson JM (1980) Application of ultrasonic pulse-echo techniques to egg albumen quality testing: preliminary report. Br Poult Sci 21:489–495

Pryor AW, Reed RDC, Richardson EG (1958) The propagation of ultrasonic waves in sols and gels. In: Stainsby G (ed) Recent advances in gelatin and glue research. Pergamon Press, London

Rahalkar RR, Gladwell N, Javanaud C, Richmond P (1986) Ultrasonic behavior of glass-filled polymer solutions. J Acoust Soc Am 80:147–156

Saggin R, Coupland JN (2002) Measurement of solid fat content by ultrasonic reflectance in model systems and chocolate. Food Res Int 35:999–1005

Saggin R, Coupland JN (2004) Shear ad longitudinal ultrasonic measurements of solid fat dispersions. J Am Oil Chem Soc 81:27–32

Santacatalina JV, Garice-Perez JV, Corona E, Benedito J (2011) Ultrasonic monitoring of lard crystallization during storage. Food Res Int 44:146–155

Saraf B, Mishra SC, Samal K (1982) Ultrasonic velocity and absorption in reconstituted powdered milk. Acustica 52:40–42

Shoh A (1988) Industrial applications of ultrasound. In: Suslick KS (ed) Ultrasound, its chemical, physical, and biological effects. VCH Publishers Inc, New York, pp 97–122

Singh AP, McClements DJ, Marangoni AG (2004) Solid fat content determination by ultrasonic velocimetry. Food Res Int 37:545–555

Ter Haar GR (1988) Biological effects of ultrasound in clinical applications. In: Suslick KS (ed) Ultrasound, its chemical, physical, and biological effects. VCH Publishers Inc, New York, pp 305–320

Ye Y, Wagh A, Martini S (2011) Using high intensity ultrasound as a tool to change the functional properties of interesterified soybean oil. J Agric Food Chem 59:10712–10722. doi:10.1021/jf202495b

Yucel U, Coupland JN (2010) Ultrasonic characterization of lactose dissolution. J Food Eng 98:28–33

Yucel U, Coupland JN (2011a) Ultrasonic characterization of lactose crystallization in gelatin gels. J Food Sci 76:E48–E54

Yucel U, Coupland JN (2011b) Ultrasonic attenuation measurements of the mixing, agglomeration, and sedimentation of sucrose crystals suspended in oil. J Am Oil Chem Soc 88:33–38

Chapter 4
Common Uses of Power Ultrasound in the Food Industry

The use of power ultrasound in the food industry has increased exponentially in the last 10 years primarily due to significant improvements in the equipment itself and the viability of implementing the technology in industrial settings. Feng et al. (2011) provide an exhaustive description of power ultrasound applications for food and bioprocessing and provide a complete explanation of the interactions between acoustic waves and food materials. Important uses of power ultrasound in food systems include its use to assist extraction processes, as an emulsification device, to induce the crystallization of food components, as a filtration aid, to change the viscosity of materials, as a defoaming agent, to help in extrusion processes, to inactivate enzymes and microbes, as a fermentation aid, to tenderize meats, and to improve heat transfer in materials. There are several excellent reviews in the literature that describe these processes in detail (Patist and Bates 2008; Chandrapala et al. 2012; Chemat et al. 2011; Rastogui 2011; Mason 1999) and therefore only a brief description of the uses of power ultrasound in food processing will be included in this chapter. Readers should refer to specific citations for further information.

4.1 Ultrasound-Assisted Extraction

Power ultrasound has been used to improve the extraction efficiency of several compounds. Acoustic waves enhance the rate and extent of mass transfer and also have a mechanical effect on cellular structure promoting solvent penetration in the cells and therefore improving the release of intracellular components. Power ultrasound has been used in a myriad of products from the extraction of oils in soybeans (Haizhou et al. 2004) to the extraction of polyphenols in coconut shells (Rodrigues et al. 2008). Detailed references describing the use of power ultrasound in the extraction of food components can be found in Rastogui (2011), Chemat et al. (2011), and Patist and Bates (2008).

S. Martini, *Sonocrystallization of Fats*, SpringerBriefs in Food, Health, and Nutrition, DOI: 10.1007/978-1-4614-7693-1_4, © Silvana Martini 2013

4.2 Emulsification

Emulsification is one of the most common uses of power ultrasound. Chandrapala et al. (2012) and Patist and Bates (2008) provide an excellent review of the uses of ultrasound to form micro-, macro-, and nano-emulsions. Power ultrasound can also be used to homogenize milk and juices, mayonnaise, and tomato ketchup (Patist and Bates 2008). The effect of power ultrasound on emulsification is driven by the collapse of cavitating bubbles near the oil/water interface and by high mixing in the system. These two events result in the formation of emulsions. As expected, the characteristics of the emulsion formed, such as droplet size and stability, are affected by sonication conditions including frequency, power level, and duration. References for use of power ultrasound to generate emulsions are included in Chandrapala et al. (2012) and Patist and Bates (2008).

4.3 Viscosity Modifier

Power ultrasound has been used to decrease the viscosity of starch-based systems such as corn, potato, tapioca, and sweet potato (Iida et al. 2008; Jambrak et al. 2010). These authors suggest that the cavitation bubbles, high pressures, and high local velocities of liquid generated through power ultrasound break the starch granules resulting in a significant decrease in viscosity. Ashokkumar's research group (2009a,b) showed a significant reduction of viscosity in sonicated whey solutions. However, Martini and Walsh (2012) and Kresic et al. (2008) showed an increase in viscosity of sonicated whey suspensions. The increase in viscosity was attributed to higher water binding capacity of the sonicated protein. Bates et al. (2006) also showed an increase in viscosity of a sonicated vegetable puree. These authors claim that ultrasound allows for an enhanced penetration of moisture in the fiber of the food material resulting in an increase in viscosity.

4.4 Defoaming

Ultrasound is used as a processing tool to defoam food products including carbonated beverages and fermented products. The ultrasound energy applied to foams breaks down the liquid film present in the foam and provides an efficient and clean alternative for breaking air bubbles. The benefits of using sonication to defoam materials over other techniques including mechanical breakers and anti-foaming agents, is that it allows for sterile defoaming and prevents chemical contamination (Patist and Bates 2008; Chemat et al. 2011).

4.5 Pasteurization

Rastogui (2011) describes the use of power ultrasound to inactivate vegetative cells and spores. Research has been performed in juices such as orange, apple, and tomato. It has also been used as a pasteurization process in dairy products, mainly milk. The effectiveness of power ultrasound in reducing microbial count results from a combination of acoustic waves and high pressures and temperatures originated during sonication. This technique has also been used to inactivate food enzymes such as peroxidases and lipoxygenases. For a complete list of references describing the use of power ultrasound in pasteurization or microbial and enzyme inactivation, refer to Rastogui (2011), Chemat et al. (2011), Condon et al. (2011); Alzamora et al. (2011); Mawson et al. (2011); and Patist and Bates (2008). The advantages of ultrasonic pasteurization include lowering flavor loss and saving energy. Pasteurization efficiency can be enhanced by combining ultrasound with heat (thermosonication), pressure (manosonicaton) or both (manothermosonication) (Chemat et al. 2011).

4.6 Sonocrystallization

In the food industry one of the main uses of sonocrystallization is for freezing foods. Power ultrasound has been used to induce the nucleation of ice and therefore increase the freezing rate of foods. Basic research on ice crystallization has been performed by Suslick (1988), Yu et al. (2012), and Chow et al. (2003, 2004, 2005). This research showed that cavitation bubbles can act as nuclei for ice crystal formation and that microstreaming and shear effects generated during sonication enhance heat and mass transfer, increasing the freezing rate (Li and Sun 2002; Zheng and Sun 2005). Some of the applied research in ultrasound-assisted freezing of foods has been performed in potatoes (Li and Sun 2002; Sun and Li 2003) and apples (Delgado et al. 2009). Russell et al. (1998) showed that power ultrasound can also be used to break ice crystals during freezing of ice cream to obtain a smoother texture.

During the last decade, power ultrasound has been used to assist the crystallization of other food materials such as edible fats. This topic will be discussed in detail in Chap. 6 and the mechanisms involved in sonocrystallization events will be discussed in detail in Chap. 5.

4.7 Other Uses of Power Ultrasound in Food Systems

Other uses of power ultrasound in the food systems include fermentation (Sinisterra 1992; Pitt and Rodd 2003; Matsuura et al. 1994), heat transfer (Kim et al. 2004; Fuente-Blanco et al. 2006), extrusion (Knorr et al. 2004; Akbari et al. 2007;

Mousavi et al. 2007), filtration (Telsonic Group 2007; Grossner et al. 2005; Muthukumaran et al. 2005; Feng et al. 2006; Kyllonen et al. 2005; Mason et al. 1996; Grossner et al. 2005), degassing (Chemat et al. 2011), depolymerization (Kassai et al. 2008; Schmid and Rommel 1939; Gronroos et al. 2004; Zuo et al. 2009; Kardos and Luche 2001; Vodenicarova et al. 2006; Drimalova et al. 2005; Wu et al. 2008; Baxter et al. 2005), cooking (Pohlman et al. 1997), drying (Garcia-Perez et al. 2009; Simal et al. 1997, 1998; Mulet et al. 2003; Jambrak et al. 2007), and changing the functional properties of proteins such as dairy and poultry (Martini et al. 2010; Martini and Walsh 2012; Gordon and Pilosof 2010; Arzeni et al. 2012; Zisu et al. 2011; Kresic et al. 2008).

In summary, power ultrasound has a wide range of applications in the food industry. Power ultrasound uses highly invasive acoustic waves that significantly change the physical and chemical properties of materials. Depending on the acoustic power and the processing conditions used, the following events induced by high intensity acoustic waves can occur separately or in combination:

- Generation of cavitation
- Generation of high shear forces
- Increase in temperatures
- Generation of free radicals
- Increase of mass transfer
- Disruption of cell material

Details regarding the specific events related to sonocrystallization will be discussed in Chap. 5.

References

Akbari Mousavi SAA, Feizi H, Madoliat R (2007) Investigations on the effects of ultrasonic vibrations in the extrusion process. J Mater Process Technol 187–188:657–661

Alzamora SM, Guerrero SN, Schenk M, Raffellini S, Lopez-Malo A (2011) Inactivation of microorganisms. In: Feng H, Barbosa-Canovas GV, Weiss J (eds) Ultrasound technologies for food and bioprocessing, 1st edn. Springer, New York, p 321–343

Arzeni C, Pérez OE, Pilosof AMR (2012) Functionality of egg white proteins as affected by high intensity ultrasound. Food Hydrocolloid 29:308–316

Ashokkumar M, Bhaskarcharya R, Kentish S, Lee J, Palmer M, Zisu B (2009a) The ultrasonic processing of dairy products – an overview. Dairy Sci Technol 90:147–168

Ashokkumar M, Lee J, Zisu B, Bhaskarcharya R, Palmer M, Kentish S (2009b) Sonication increases the heat stability of whey proteins. J Dairy Sci 92:5353–5356

Bates DM, Bagnall WA, Bridges MW (2006) Method of treatment of vegetable matter with ultrasonic energy. US Patent 20,060,110,503

Baxter S, Zivanovic S, Weiss J (2005) Molecular weight and degree of acetylation of high-intensity ultrasonicated chitosan. Food Hydrocolloid 19:821–830

Chandrapala J, Oliver C, Kentish S, Ashokkumar M (2012) Ultrasonics in food processing. Ultrason Sonochem 19:975–983

Chemat F, Huma Z, Khan MK (2011) Applications of ultrasound in food technology: processing, preservation and extraction. Ultrason Sonochem 18:813–835

Chow R, Blindt R, Chivers R, Povey M (2003) The sonocrystallisation of ice in sucrose solutions: primary and secondary nucleation. Ultrasonics 41:595–604

Chow R, Blindt R, Kamp A, Grocutt P, Chivers R (2004) The microscopic visualisation of the sonocrystallisation of ice using a novel ultrasonic cold stage. Ultrason Sonochem 11:245–250

Chow R, Blindt R, Chivers R, Povey M (2005) A study on the primary and secondary nucleation of ice by power ultrasound. Ultrasonics 43:227–230

Condon S, Mañas P, Cebrian G (2011) Manothermosonication for microbial inactivation. In: Feng H, Barbosa-Canovas GV, Weiss J (eds) Ultrasound technologies for food and bioprocessing, 1st edn. Springer, New York, pp 287–319

Delgado AE, Zheng L, Sun DW (2009) Influence of ultrasound in freezing rate of immersion-frozen apples. Food Bioproc Technol 2:263–270

Drimalova A, Velebny V, Sasinkova V, Hromadkova Z, Ebrigerova A (2005) Degradation of hyaluronan by ultrasonication in comparison to microwave and conventional heating. Carbohydr Polym 61:420–426

Feng D, van Deventer JSJ, Aldrich C (2006) Ultrasonic defouling of reverse osmosis membranes used to treat wastewater effluents. Sep Purif Technol 50:318–323

Feng H, Barbosa-Canovas GV, Weiss J (2011) Ultrasound technologies for food and bioprocessing. Feng H, Barbosa-Canovas GV, Weiss J (eds) Springer, New York

Fuente-Blanco S, Riera-Franco de Sarabia E, Acosta-Aparicio VM, Blanco–Blanco A, Gallego-Juárez JA (2006) Food drying process by power ultrasound. Ultrasonics 44:e523–e527

Garcia-Perez JV, Carcel JA, Riera E, Mulet A (2009) Influence of the applied acoustic energy on the drying of carrots and lemon peel. Drying Technol 27:281–287

Gordon L, Pilosof AMR (2010) Application of high-intensity ultrasounds to control the size of whey proteins articles. Food Biophys 5:203–210

Gronroos A, Pirkonen P, Ruppert O (2004) Ultrasonic depolymerization of aqueous carboxy-methylcellulose. Ultrason Sonochem 11:9–12

Grossner MT, Belovich JM, Feke DL (2005) Transport analysis and model for the performance of an ultrasonically enhanced filtration process. Chem Eng Sci 60:3233–3238

Haizhou L, Pordesimo L, Weiss J (2004) High intensity ultrasound-assisted extraction of oil from soybeans. Food Res Int 37:731–738

Iida Y, Tuziuti T, Yasui K, Towata A, Kozuka T (2008) Control of viscosity in starch and polysaccharide solutions with ultrasound after gelatinization. Innov Food Sci Emerg Technol 9:140–146

Jambrak AR, Mason TJ, Paniwnyk L, Lelas V (2007) Accelerated drying of button mushrooms, brussels sprouts and cauliflower by applying power ultrasound and its rehydration properties. J Food Eng 81:88–97

Jambrak A, Herceg Z, Subaric DD, Babic J, Brncic S, Bosiljkov T, Cvek D, Tripalo B, Gelo J (2010) Ultrasound effect on physical properties of corn starch. Carbohydr Polym 79:91–100

Kardos N, Luche J-L (2001) Sonochemistry of carbohydrate compounds. Carbohydr Res 332:115–131

Kasaai MR, Arul J, Charlet G (2008) Fragmentation of chitosan by ultrasonic irradiation. Ultrason Sonochem 15:1001–1008

Kim HY, Kim YG, Kang BH (2004) Enhancement of natural convection and pool boiling heat transfer via ultrasonic vibration. Int J Heat Mass Tran 47:2831–2840

Knorr D, Zenker M, Heniz V, Lee D-U (2004) Applications and potential of ultrasonics in food processing. Trend Food Sci Technol 15:261–266

Kresic G, Lelas V, Jambrak AR, Herceg Z, Brncic SR (2008) Influence of novel processing technologies on the rheological and thermophysical properties of whey proteins. J Food Eng 87:64–73

Kyllonen HM, Pirkonen P, Nystrom M (2005) Membrane filtration enhanced by ultrasound: a review. Desalination 181:319–335

Li B, Sun DW (2002) Effect of power ultrasound in freezing rate during immersion freezing. J Food Eng 55:277–282

Martini S, Walsh MK (2012) Sensory characteristics and functionality of sonicated whey. Food Res Int 49:694–701

Martini S, Potter R, Walsh MK (2010) Optimizing the use of high intensity ultrasound to decrease turbidity in whey protein suspensions. Food Res Int 43:2444–2451

Mason J, Paniwynyk L, Lorimer P (1996) The use of ultrasound in food technology. Ultrason Sonochem 3:S253–S260

Mason TJ (1999) Sonochemistry: current uses and future prospects in the chemical and processing industries. Philos T R Soc A 357:355–369

Matsuura K, Hirotsune M, Nunokawa Y, Satoh M, Honda K (1994) Acceleration of cell growth and ester formation by ultrasonic wave irradiation. J Ferment Bioeng 77:36–40

Mawson R, Gamage M, Terefe NS, Knoerzer K (2011) Ultrasound in enzyme activation and inactivation. In: Feng H, Barbosa-Canovas GV, Weiss J (eds) Ultrasound technologies for food and bioprocessing, 1st edn. Springer, New York, pp 369–404

Mousavi SAAA, Feizi H, Madoliat R (2007) Investigations on the effect of ultrasonic vibrations in the extrusion process. J Mater Process Technol 187–188:657–661

Mulet A, Carcel JA, Sanjuan N, Bon J (2003) New food drying technologies – use of ultrasound. Food Sci Technol Int 9:215–221

Muthukumaran S, Kentish SE, Ashokkumar M, Stevens GW (2005) Mechanisms for the ultrasonic enhancement of dairy whey ultrafiltration. J Membr Sci 258:106–114

Patist A, Bates D (2008) Ultrasonic innovations in the food industry: from the laboratory to commercial production. Innov Food Sci Emerg Technol 9:147–154

Pitt WG, Rodd A (2003) Ultrasound increases the rate of bacterial growth. Biotechnol Prog 19:1030–1044

Pohlman FW, Dikeman ME, Zayas JF, Unruh JA (1997) Effects of ultrasound and convection cooking to different end point temperatures on cooking characteristics, shear force and sensory properties, composition and microscopic morphology of beef longissimus and pectoralis muscles. J Anim Sci 75:386–401

Rastogui NK (2011) Opportunity and challenges in application of ultrasound in food processing. Crys Rev Food Sci Nutr 51:705–722

Rodrigues S, Pinto GAS, Fernandes FAN (2008) Optimization of ultrasound extraction of phenolic compounds from coconut (Cocos nucifera) shell powder by response surface methodology. Ultrason Sonochem 15:95–100

Russell AB, Cheney PE, Wantling SD (1998) Influence of freezing conditions on ice crystallization in ice cream. J Food Eng 39:179–191

Schmid G, Rommel O (1939) Rupture of macromolecules with ultrasound. Z Phys Chem A 185:97–139

Simal S, De Mirabo FB, Deya E, Rossello C (1997) A simple model to predict the mass transfers in osmotic dehydration. Lebensm Untersuch Forsch 204:210–214

Simal S, Benedito J, Sanchez ES, Rossello C (1998) Use of ultrasound to increase mass transportrates during osmotic dehydration. J Food Eng 36:323–336

Sinisterra JV (1992) Application of ultrasound to biotechnology: an overview. Ultrasonics 30:180–185

Sun DW, Li B (2003) Microstructural change of potato tissues frozen by ultrasound-assisted immersion freezing. J Food Eng 57:337–345

Suslick KS (1988) Homogeneous sonochemistry. In: Suslick KS (ed) Ultrasound: its chemical, physical, and biological effects, 1st edn. VCH Publishers Inc, New York, pp 123–163

Telsonic Group (2007) Ultrasonic screening technology, Bronschhofen, Switzerland. http://www.telsonic.com

Vodenicarova M, Drimalova G, Hromadkova Z, Malovikova Z, Ebringerova A (2006) Xyloglucan degradation using different radiation sources: a comparative study. Ultrason Sonochem 13:157–164

Wu T, Zivanovic S, Hayes DG, Weiss J (2008) Efficient reduction of chitosan molecular weight by high-intensity ultrasound: underlying mechanism and effect of process parameters. J Agric Food Chem 56:5112–5119

Yu D, Liu B, Wang B (2012) The effect of ultrasonic waves on the nucleation of pure water and degassed water. Ultrason Sonochem 19:459–463

Zheng L, Sun DW (2005) Ultrasonic assistance of food freezing. In: Sun DW (ed) Emerging technologies for food processing, 1st edn. Elsevier, London, p 603–627

Zisu B, Lee J, Chandrapala J, Bhaskaracharya R, Palmer M, Kentish S, Ashokkumar M (2011) Effect of ultrasound on the physical and functional properties of reconstituted whey protein powders. J Dairy Res 78:226–232

Zuo JY, Knoerzer K, Mawson R, Kentish S, Ashokkumar M (2009) The pasting properties of sonicated waxy rise starch suspensions. Ultrason Sonochem 16:462–468

Chapter 5
Mechanisms Involved in Sonocrystallization

The focus of this Brief is to discuss the use of power ultrasound to induce the crystallization of lipids (sonocrystallization of fats). When high power acoustic waves travel through a media, changes in localized pressures generated by acoustic waves induce the formation of cavities. Cavities, and the phenomena associated with their formation, are responsible for the destructive effect of high power acoustic waves. The phenomena of creating cavities with acoustic waves is called acoustic cavitation.

5.1 Acoustic Cavitation

Cavitation as defined by Leighton (1994) is a phenomenon occurring *"whenever a new surface, or cavity, is created within the body of a liquid, a cavity being defined as any bounded volume, whether empty or containing gas or vapor, with at least part of the boundary being liquid."* Cavitation can occur in any liquid from water to molten metal; and lipids are no exception. Common cavitation events include underwater explosions, effervescence, and the boiling of liquid. Cavitation induced by acoustic waves is called *acoustic cavitation*. Cavitation can occur in a pure liquid in the absence of any inhomogeneity such as solid particles (*homogeneous* cavitation), or in a liquid with microscopic impurities such as microbubbles, small particles of dust, or gas pockets (*heterogeneous* cavitation). A minimum amount of acoustic power must be delivered to induce homogeneous or heterogeneous cavitation. Several studies have focused on measuring acoustic thresholds for cavitation to occur. *Acoustic threshold* is defined as the minimum power (or pressure amplitude) needed to generate cavitation and is a function of the characteristics of the media and on the acoustic frequency used. Factors that affect cavitation thresholds include surface tension, tensile strength of the material, temperature, and number and size of contaminants such as dirt (Atchley and Crum 1988). The formation of cavities is induced when the acoustic pressure of the wave is greater than the static pressure in the system. This will generate zones of negative pressure in the rarefaction phase (Chap. 2) that will place the material

S. Martini, *Sonocrystallization of Fats*, SpringerBriefs in Food, Health, and Nutrition, DOI: 10.1007/978-1-4614-7693-1_5, © Silvana Martini 2013

into tension. This tension in the material will eventually result in the formation of cavities. Impurities create "weaknesses" in the material and in this case lower acoustic power levels are needed to generate cavitation. Therefore, the cavitation threshold for an impure material is lower than the cavitation threshold for a pure material (Leighton 1994).

Cavitation research includes not only the study of bubble or cavity formation but also the life and dynamics of bubbles created including bubble growth, dissolution, oscillation, and collapse. According to the life cycle or dynamics of bubbles formed, cavitation can be classified into two types: *inertial* (or transient) and *non-inertial* (or stable) cavitation. During high intensity sonication, bubbles or cavities are formed and oscillate around their equilibrium position with little change in bubble size over time. We refer to this as non-inertial or stable cavitation, where the amplitude of the oscillation is small compared to the bubble equilibrium radius. However, under specific conditions, these bubbles can grow in size as a function of time and they will eventually collapse. This is called inertial cavitation. This collapse is violent and it is associated with high localized temperatures, pressures, and shear forces. Inertial cavitation occurs during the compression phase of bubble life where the compression is so chaotic that bubbles become unstable and collapse (Frizzel 1988). Depending on the acoustic power level used the formation of stable cavitation, inertial cavitation, or a combination of both is possible. It is important to understand that these events are not isolated and that interactions occur among the cavities or bubbles formed. For example, non-inertial bubbles can eventually grow in size and generate inertial cavities while inertial bubbles can form stable cavities after they collapse.

The formation and presence of acoustic cavities are responsible for the several physicochemical changes observed during high power sonication. For example, stable microbubbles generated during sonication can act as nuclei to induce crystallization while high shear forces and high temperatures associated with the collapse of inertial cavities are responsible for induction of chemical reactions and increased heat and mass transfer.

5.2 Rectified Diffusion

As described above bubbles can grow in response to acoustic waves. Bubble growth during stable cavitation is mainly driven by a phenomenon called *rectified diffusion*. When the bubble oscillates there is an exchange of gas between the bubble and the media during each acoustic cycle. While bubbles contract the gas pressure inside the bubble is greater than the equilibrium value and therefore gas diffuses from the inside the bubble into the surrounding media or liquid. Conversely, when the bubble expands, gas pressure inside the bubble is lower than the value at equilibrium and therefore gas diffuses from the liquid into the bubble. Since the area of the bubble is bigger during the expansion cycle, more gas enters into the bubble during the expansion phase than the gas that diffuses from the

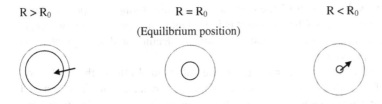

Fig. 5.1 Schematic representation of the area and shell effect responsible for rectified diffusion. R = instantaneous bubble radius; R_0 = equilibrium bubble radius; $\rightarrow$ = Gas flow; ... = liquid shell. Reprinted with modifications from Leighton TG (1994) with permission from Elsevier

bubble during the contraction phase. Therefore, over a number of cycles of the acoustic wave there is a net positive flow of gas to the interior of the bubble resulting in bubble growth. This is called the *area effect* of rectified diffusion. In addition to the area effect, *shell effect* can also contribute to rectified diffusion. The volume of the spherical shell surrounding the bubble changes during oscillation resulting in a gradient in gas concentration that affects gas diffusion between the bubble and the surrounding liquid. During the expansion phase gas concentration in the wall of the bubble is lower than gas concentration in the bubble at equilibrium. In addition, the shell is smaller or contracted compared to the shell at equilibrium. These two events result in a higher gradient of gas across the shell resulting in gas diffusion into the bubble. Similarly, when the bubble contracts, the shell expands and the gas concentration in the wall of the bubble increases. Since the shell is thicker than it is in equilibrium position the gas gradient is lower resulting in slower diffusion of the gas from the bubble wall to the bulk of the liquid. The area and shell effect in combination are responsible for the rectified diffusion that results in a net flow of gas into the bubble with a consequent increase in bubble size (Leighton 1994). Figure 5.1 shows a schematic representation of the events that occur during rectified diffusion. In this figure, the wall of the bubble is indicated with a black line, while the shell of liquid that surrounds the bubble is denoted with a dotted line. When the bubble expands ($R > R_0$) the pressure of gas inside the bubble decreases and there is a net flow of gas to the interior of the bubble due to the area and shell effect. Similarly, when the bubble is compressed, there is an increase in pressure inside the bubble which forces gas to diffuse from the interior of the bubble to the surrounding liquid.

5.3 Sonocrystallization

The effect of power ultrasound on crystallization of organic and inorganic materials has been studied for the past 70 years. Luque de Castro and Priego-Capote (2007) provide a comprehensive description of the effects of power ultrasound on crystallization. A description of the effect of power ultrasound on lipid crystallization will be presented in Chap. 6, but it is important to first understand the effects of power ultrasound in organic and inorganic molecules and extrapolate this knowledge to lipid systems.

The scientific community agrees that power ultrasound affects the nucleation and growth of crystallizing molecules through several mechanisms. The first and most important phenomenon is the formation of acoustic cavities. The formation of acoustic cavities is associated with high pressures (100–5,000 atm), high shear forces, and high temperatures (5,000–10,000 K) that lead to induction in crystallization (Gogate and Pandit 2011). These events generate rapid cooling rates (in the order of 10^7–10^{10} K/s), a reduction in the crystallization temperature, and a lowering of the excitation energy barriers for nucleation (Luque de Castro and Priego-Capote 2007). Notwithstanding significant evidence showing that the number of cavitating entities is related to nucleation events, more research is required to understand the mechanisms for these events. In addition to inducing nucleation and crystal growth, power ultrasound is used to generate materials with small crystal sizes and to improve product quality and processes reproducibility.

Power ultrasound has been used to crystallize several materials including potassium sulfate (Lyczko et al. 2002), potash alum (Amara et al. 2001, 2004), 7-amino-3-desacetoxy cephalosporanic acid (7-ACDA) (Li et al. 2006), lactose (Dhumal et al. 2008; Bund and Pandit 2007a, b; Patel and Murthy 2009), and ice (Chow et al. 2003, 2004, 2005). For detailed information refer to Gogate and Pandit (2011) and Luque de Castro and Priego-Capote (2007).

Independent of the material used, the effect of power ultrasound on crystallization of organic and inorganic molecules depends on several factors including (Luque de Castro and Priego-Capote 2007):

- *Acoustic frequency*: power ultrasound can affect crystallization when used at low frequencies of the order of 15–30 kHz (Li et al. 2003). The higher the frequency the lower the acoustic power and therefore fewer effects on crystallization are expected.
- *Acoustic intensity, power and horn tip size*: An increase in acoustic intensity, power, and tip size increases the crystallization rate (Nishida 2004; Li-yun et al. 2005; Li et al. 2003; Amara et al. 2001).
- *Immersion depth*. The immersion depth affects the flow pattern of the liquid and therefore the crystallization behavior. Immersion depth used depends on type of tip and sonicated material (Nishida 2004).
- *Volume of sonicated solution*. For a constant power level, the acoustic waves have a lower penetration capacity when used in larger volumes (Amara et al. 2001). Therefore, the greater the volume of the solution, the smaller the effect of ultrasound.
- *Duration of the sonication*. Crystal sizes decrease with increased sonication time. Manipulation of sonication time is useful to control and tailor crystal size. Li et al. 2003; Anderson et al. 1995 demonstrated that a combination of continuous and pulsed sonication can be used to control crystal size.

In short, ultrasound process parameters including power level, sonication time, tip size, can be adjusted to control crystallization processes. The effect of these parameters on the sonocrystallization of fats will be discussed in Chap. 6.

References

Amara N, Ratsimba B, Wilhem AM, Delmas H (2001) Crystallization of potash alum: effect of power ultrasound. Ultrason Sonochem 8:265–270

Amara N, Ratsimba B, Wilhem AM, Delmas H (2004) Growth rate of potash alum crystals: comparison of silent and ultrasonic conditions. Ultrason Sonochem 11:17–21

Anderson HW, Carbery JB, Staunton HF, Sutradhar BC (1995) S Patent 5 471 001

Atchley AA, Crum LA (1988) Acoustic cavitation and bubble dynamics. In: Suslick KS (ed) Ultrasound: its chemical, physical and biological effects, 1st edn. VCH Publishers Inc, New York, pp 1–64

Bund RK, Pandit AB (2007a) Rapid lactose recovery from paneer whey using sonocrystallization: a process optimization. Chem Eng Process 46:846–850

Bund RK, Pandit AB (2007b) Rapid lactose recovery from buffalo whey by use of "antisolvent" ethanol. J Food Eng 82:333–341

Chow R, Blindt R, Chivers R, Povey M (2003) The sonocrystallisation of ice in sucrose solutions: primary and secondary nucleation. Ultrasonics 41:595–604

Chow R, Blindt R, Kamp A, Grocutt P, Chivers R (2004) The microscopic visualisation of the sonocrystallisation of ice using a novel ultrasonic cold stage. Ultrason Sonochem 11:245–250

Chow R, Blindt R, Chivers R, Povey M (2005) A study on the primary and secondary nucleation of ice by power ultrasound. Ultrasonics 43:227–230

Dhumal RS, Biradar SV, Paradkar AR, York P (2008) Ultrasound assisted engineering of lactose crystals. Pharm Res 25:2835–2844

Frizzel LA (1988) Biological effects of acoustic cavitation. In: Suslick KS (ed) Ultrasound: its chemical, physical and biological effects, 1st edn. VCH Publishers Inc, New York, pp 287–303

Gogate PR, Pandit AB (2011) Sonocrystallization and its application in food and bioprocessing. In: Feng H, Barbosa-Canovas GV, and Weiss J (eds) Ultrasound technologies for food and bioprocessing, 1st edn. Springer, New York, pp 467–493

Leighton TG (1994) The forced bubble. In: Leighton TG (ed) The acoustic bubble, 1st edn. Academic Press, New York, pp 379–381

Li H, Wang J, Bao Y, Guo Z, Zhang M (2003) Rapid sonocrystallization in the salting-out process. J Cryst Growth 247:192–198

Li H, Li H, Guo Z, Liu Y (2006) The application of power ultrasound to reaction crystallization. Ultrason Sonochem 13:359–363

Li-yun C, Chuan-bo Z, Jian-feng H (2005) Influence of temperature, [Ca2+], Ca/P ratio and ultrasonic power on the crystallinity and morphology of hydroxyapatite nanoparticles prepared with a novel ultrasonic precipitation method. Mat Lett 59:1902–1906

Luque de Castro MD, Priego-Capote F (2007) Ultrasound-assisted crystallization (sonocrystallization). Ultrason Sonochem 14:717–724

Lyczko N, Espitalier F, Louisnard O, Schwartzentruber J (2002) Effect of ultrasound on the induction time and the metastable zone widths of potassium sulphate. Chem Eng J 86:233–241

Nishida I (2004) Precipitation of calcium carbonate by ultrasonic irradiation. Ultrason Sonochem 11:423–428

Patel SR, Murthy ZVP (2009) Ultrasound assisted crystallization for the recovery of lactose in an anti-solvent acetone. Crystal Res Technol 44:496–889

Chapter 6
Sonocrystallization of Fats

The previous chapter described phenomena responsible for inducing crystallization with acoustic waves and also provided several examples of power ultrasound process parameters and their effectiveness. Acoustic waves have been used to crystallize organic and inorganic compounds for the past 70 years; however, there has been surprisingly little research on food-related materials. Some of the initial food-related work was performed in ice crystallization to improve freezing operations of food systems (Chow et al. 2003, 2004, 2005; Sun and Li 2003; Li and Sun 2002), followed by lactose and sucrose crystallization experiments (Patel and Murthy 2009; Dhumal et al. 2008; Bund and Pandit 2007a, b). The first body of research in sonocrystallization of lipids was reported in early 2000. Sato's research group pioneered the study of using power ultrasound to crystallize fats (Higaki et al. 2001; Ueno et al. 2003, 2002) followed by further work by Patrick et al. (2004) and Martini et al. (2008, 2012), Suzuki et al. (2010), Ye et al. (2011). During this period, several patents were issued on the use of power ultrasound in lipid fractionation (Arends et al. 2003) and chocolate manufacture (Baxter et al. 1997a and b).

This chapter will describe the use of power ultrasound to induce crystallization of fats and will include specific experimental detail needed to achieve these results.

6.1 Anhydrous Milk Fat

Power ultrasound was applied to Anhydrous Milk Fat (AMF) in a laboratory setting and the effect of acoustic waves on the crystallization behavior and functional properties of the AMF were tested (Martini et al. 2008; Suzuki et al. 2010).

6.1.1 Experimental Conditions

AMF was melted at 80 °C for 30 min prior to the crystallization experiments. The molten sample (100 g) was placed in a double-walled crystallization cell which

S. Martini, *Sonocrystallization of Fats*, SpringerBriefs in Food, Health, and Nutrition, 41
DOI: 10.1007/978-1-4614-7693-1_6, © Silvana Martini 2013

Fig. 6.1 Double-walled crystallization cell used to crystallize AMF at different T_c. Ultrasound tip (or probe) is immersed in the sample to generate the acoustic waves and to measure sample temperature, respectively. A magnetic stirrer was used at 200 rom to ensure efficient heat transfer during the crystallization experiment. Reprinted from Martini et al. (2008) with kind permission from Springer Science and Business Media

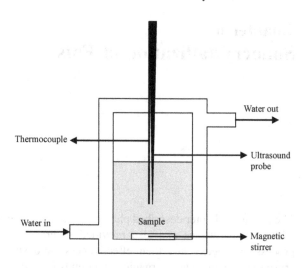

was maintained at a constant temperature using an external water bath. Figure 6.1 shows a schematic representation of the crystallization cell used in this experiment. As seen in this figure the ultrasound tip was immersed in the sample and a thermocouple was attached to it. This allows for accurate measurement of the sample temperature during crystallization and sonication. A magnetic stirrer was used at 200 rpm to ensure efficient heat transfer during the crystallization experiment.

Once the hot sample (~ 60 °C) was placed in the crystallization cell the temperature in the bulk lipid decreased exponentially at an average cooling rate of 10 °C/min to reach crystallization temperature (T_c). The effect of power ultrasound on crystallization of AMF was tested at different T_c values ($T_c = 22, 24, 26, 28,$ and 30 °C) and the presence and growth of crystals was followed using polarized light microscopy (PLM, Olympus BX41, Tokyo, Japan) for a period of 90 min. Induction times of crystallization were recorded as the time when the first crystals were observed by the naked eye and corroborated by PLM. Power ultrasound was applied to the sample 10 min after the sample was placed in the crystallization cell, which corresponds to the time when the sample reached crystallization temperature. A Misonix 3000 (Misonix Inc., NY, USA) sonicator with a maximum power output of 300 W and a frequency of 20 kHz was used to sonicate the sample for 10 s using 50 W of electrical power and a 3.2 mm diameter tip with a vibration amplitude of 168 μm. After maintaining the samples in the crystallization cell for 90 min the viscosity was measured using a Brookfield Viscometer (Model DV-II+). In addition, hardness was evaluated after tempering the samples at 5 °C for 24 h using a TA-XT plus Texture Analyzer (Texture Technologies, Scarsdale, NY).

6.1.2 Results and Discussion: Effect of Sonication on the Crystallization of AMF at Different Supercoolings

The T_c tested in this experiment ($T_c = 22, 24, 26, 28,$ and 30 °C) were chosen to generate different degrees of supercoolings to affect the crystallization behavior of the samples. Supercooling (ΔT) is defined as the difference between the melting point (T_m) of the sample and the crystallization temperature (T_c). Slower crystallization and longer induction times of crystallization are expected when the sample is crystallized at higher T_c (lower supercoolings), and faster crystallization with shorter induction times are expected at lower T_c (higher supercoolings). The melting point of the AMF sample used in these experiments was 32.4 ± 0.6 °C and therefore, the range of supercoolings ($\Delta T = T_m - T_c$) tested was between 10.4 and 2.4 °C. When power ultrasound was applied to AMF a significant ($\alpha = 0.05$) reduction in the induction time of crystallization was observed, especially for samples crystallized at intermediate supercoolings (26 and 28 °C). Table 6.1 shows the induction times of crystallization for sonicated and non-sonicated samples crystallized at different temperatures.

AMF crystallized without the use of power ultrasound had induction times of 20.0 ± 0.0 and 35.0 ± 2.8 min at 26 and 28 °C, respectively. These values decreased to 16.0 ± 1.4 and 20.5 ± 2.1 °C when power ultrasound was used during the crystallization process. No effect of sonication was observed on the induction time of crystallization for samples crystallized at 22 and 24 °C, with induction times of approximately 14.5 and 18.5, respectively. In fact, a slight increase in the induction time of crystallization was observed in samples crystallized at 22 °C with values of 13.5 ± 2.1 min for AMF crystallized without power ultrasound and 15.5 ± 0.7 min for AMF crystallized with power ultrasound application. Interestingly, when AMF was crystallized at very low supercoolings ($T_c = 30$ °C) no decrease in induction time of crystallization was observed. The lower induction times obtained for sonicated samples crystallized at intermediate supercoolings ($T_c = 26$ and 28 °C) suggest that crystallization occurs sooner under these conditions. In this case, it is likely that primary and/or secondary

Table 6.1 Induction times of crystallization (min) determined by the naked eye and confirmed by polarized light microscopy for AMF samples crystallized at different crystallization temperatures with (with HIU) and without sonication (without HIU)

Induction times (min)		
T_c (°C)	AMF without HIU	AMF with HIU
22	13.5 ± 2.1	15.5 ± 0.7
24	19.0 ± 1.4	18.0 ± 0.0
26	20.0 ± 0.0	16.0 ± 1.4
28	35.0 ± 2.8	20.5 ± 2.1
30	33.5 ± 0.7	30.5 ± 4.9

Mean values and standard deviations are reported (Martini et al. 2008)

nucleation events are induced by high intensity acoustic waves resulting in earlier onset of crystallization. Crystallization experiments described were performed isothermally. Induction times, as defined in this study, are above 10 min, which is the time that takes for the sample to reach T_c. Therefore, the onset of crystallization occurred under isothermal conditions for all the supercoolings tested in this experimental design. Consequently, no crystals were present in the media when power ultrasound was applied at 10 min into the crystallization process suggesting that high intensity acoustic waves induce primary nucleation in the system. The induction of crystallization generated by power ultrasound is evident from the crystals shown in Fig. 6.2. This figure shows PLM micrographs of AMF crystals obtained as a function of time for samples crystallized at 28 °C with and without the use of power ultrasound (Martini et al. 2008). The first column in this figure shows crystals obtained during the crystallization of AMF without the use of power ultrasound at different time points (rows). The second column in the figure shows crystals obtained during the sonocrystallization of AMF. It is clear from Fig. 6.2 that power ultrasound induces crystallization of AMF; more crystals are observed for each point in time for sonicated samples. This figure also suggests that crystal growth is promoted by sonication since more crystals are observed even at longer times during the crystallization process (for example at $t = 80$ min).

Fig. 6.2 PLM micrographs of AMF crystals as a function of crystallization time when crystallized at 28 °C without power ultrasound (*first column*) and with power ultrasound (*second column*). *Rows* in each column represent the times when samples were taken from the crystallization cell. Reprinted from Martini et al. 2008 with kind permission from Springer Science and Business Media

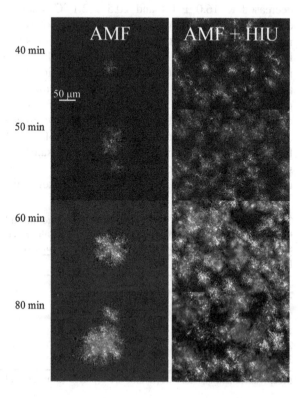

As noted above, no effect of power ultrasound on induction of crystallization was observed for AMF samples crystallized at lower supercoolings (30 °C). Recall that power ultrasound is applied at 10 min into the crystallization process, and that AMF starts crystallizing at approximately 35 min when crystallized at $T_c = 30$ °C without the use of power ultrasound. It is likely that the effect of sonication is lost when applied at such early stages in the crystallization process. At this low supercooling the system is in a metastable zone and although the thermodynamic force for crystallization is reached the system needs some time to crystallize. When power ultrasound is applied at early stages in the crystallization process the effect of sonication is lost before molecular entities can find the right conformation to form new nuclei. This hypothesis was tested by applying power ultrasound at low supercoolings but closer to the onset of crystallization. In this case an induction in crystallization was observed when power ultrasound was applied after the first crystals were observed in the media and when lower power levels of 30 and 5 W (tip vibration amplitude of 120–124 μm) and shorter durations (5 s) were used (Martini et al. 2008).

The crystallization behavior of lipids affects physical and functional properties of the final crystal network formed. Therefore, by controlling crystallization processes it is possible to manipulate the properties of fats. These properties include melting behavior, hardness, viscosity, and solid fat content, among others. We expect therefore, that the effect of power ultrasound on crystallization of fats will affect some or all of these properties. Power ultrasound can be used not only to induce primary and secondary crystallization but also to generate smaller crystals. Martini et al. (2008) measured the effect of sonication on the viscosity and hardness of sonicated AMF. In general, when power ultrasound induced the crystallization of AMF higher viscosity values were obtained (Martini et al. 2008). In addition, a significant increase ($\alpha = 0.05$) in the hardness of sonicated AMF samples was observed, especially when crystallized at intermediate temperatures (24–28 °C). Even though smaller crystals and harder materials were generated in AMF crystallized with power ultrasound, significant differences ($\alpha = 0.05$) in the melting enthalpies were observed only for samples crystallized at 28 °C (Suzuki et al. 2010). Melting enthalpies can be used to quantify the amount of crystalline material in the sample. These results suggest that the harder texture observed in AMF samples crystallized in the presence of power ultrasound at 24–26 °C is mainly due to smaller crystals present in the material and not to the presence of more crystals in the network. The harder fat network obtained at 28 °C is due to a combination of factors such as the presence of smaller crystals and more crystallized material as seen by significantly higher melting enthalpies.

High intensity acoustic waves generated by power ultrasound techniques induce cavitation events that influence the crystallization of lipids. As discussed in Chap. 5, inertial and non-inertial cavities can be formed in the media in response to high intensity acoustic waves. Inertial cavities or bubbles are formed and then collapse in a violent manner generating high shear forces and localized high temperatures and pressures in the media. It is very likely that inertial and non-inertial microbubbles are responsible for the induction of primary nucleation since they can act as new nucleation sites. In addition, the high shear forces resulting from the violent collapse of the inertial bubbles might be responsible for the induction of secondary nucleation and crystal growth observed especially when power ultrasound is applied in the presence of a small number of crystals.

The lack of induction in crystallization and the delay in crystallization observed when power ultrasound is applied at high supercoolings (22 °C) can be explained in the context of the bubbles dynamics described above. At this low temperature more power is required to generate cavities given the increased viscosity in the oil. Moreover, the limited number of bubbles formed during sonication may be non-inertial in nature since the higher viscosity acts to stabilize the bubbles and delay their collapse. Sonication is also associated with a slight increase in sample temperature (0.5–1 °C) which might partially melt the recently formed crystals and delay crystallization. Conversely, when AMF is sonicated at 30 °C bubbles formed during the sonication process dissolve easily in the less viscous oil. When power ultrasound is applied early in the crystallization process in low supercooled materials, cavities are formed before the appropriate molecular conformation can be attained to form a nuclei. Cavities therefore are not able to interact with nuclei and no effect of acoustic waves on the crystallization behavior of the material is observed.

6.2 Palm Kernel Oil and Palm Oil

Patrick et al. (2004) reported a significant effect on palm oil crystallization using power ultrasound. In addition, Suzuki et al. (2010) studied the effect of power ultrasound on palm kernel oil. Details regarding the experimental conditions used by these researchers and the results obtained are discussed below.

6.2.1 Experimental Conditions

Palm oil: Patrick et al. (2004) used ring transducers instead of an immersed tip to generate a radial uniform acoustic field in the crystallization cell and they used vertical blades to stir the sample. Palm oil (PO) was crystallized under non-isothermal conditions at a slow cooling rate (0.1 °C/min). These authors applied power ultrasound when the sample reached 45 °C and stopped the sonication when the sample reached its final temperature of 23 °C. The power levels used by this group were 30, 35, 40, and 45 dB.

Palm kernel oil: Palm kernel oil (PKO) with a melting point of 32.9 ± 0.3 °C was crystallized as described in Sect. 6.1.1 at a $T_c = 30$ °C. Power ultrasound was applied to PKO samples for 10 s using a 3.2 mm diameter microtip at an acoustic power of 58 W using a Misonix 3000 sonicator (Misonix Inc., NY, USA). Two different conditions were used: (a) power ultrasound was applied when the sample reached crystallization temperature (T_c condition) and (b) power ultrasound was applied when the first crystals were observed in the sample by the naked eye (f_c condition). The progress of crystallization was followed using PLM. Samples were kept in the crystallization cell for 90 min to ensure that the crystallization process reached equilibrium and at this time the melting profile of the samples was measured using differential scanning calorimetry (DSC, DSC-2910, TA Instruments, New Castle, DE). The melting enthalpy (ΔH), and the onset (T_{on}) and peak (T_p) melting temperatures were used to characterize the melting profile of the system. The hardness of the material was measured after tempering the samples for 24 h at 5 °C using a double compression test with a TA-XT plus Texture Analyzer (Texture Technologies, Scarsdale, N.Y., U.S.A.).

6.2.2 Results and Discussion

Effect of power ultrasound on PO crystallization: Patrick et al. (2004) report that power ultrasound applied at intensity values below the cavitation threshold for PO (41 dB) induced the crystallization of PO and generated smaller crystals. In addition, a network of small lipid crystals was observed in the sonicated samples while big clusters of crystals were observed in the non-sonicated ones. Interestingly, these authors report that at intensities above the cavitation threshold (intensities of 45 dB) ultrasound did not affect crystal morphology of PO but slightly decreased the induction time of crystallization from 6.2 min in the samples crystallized without the use of power ultrasound to 4.8 min for the samples crystallized with power ultrasound at the highest power level tested (45 dB). These authors also reported an increase in crystallization temperature from 26 °C in non-sonicated samples to 33 °C in samples sonicated using the highest power setting (45 dB).

Effect of power ultrasound on PKO crystallization: The effect of power ultrasound on PKO crystallization was studied by Suzuki et al. (2010). The authors evaluated the effect of applying power ultrasound at different points in the crystallization process. PKO was crystallized under isothermal conditions at 30 °C as described in Sect. 6.2.1 either when the sample reached crystallization temperature (T_c condition) or at the onset of crystallization (f_c condition). The first column of Fig. 6.3 shows crystal morphology of PKO samples crystallized without power ultrasound and the second and third columns of the figure show crystals obtained when PKO is crystallized with power ultrasound under T_c and f_c conditions, respectively. When power ultrasound was applied at T_c, a slight induction in crystallization was observed (Fig. 6.3) and this effect was more significant when power ultrasound was applied at f_c. The presence of more crystals at a specific time point during early stages of crystallization (Fig. 6.3, columns 2, 3) indicates that power ultrasound induced the onset of crystallization. Similarly, the higher number of crystals observed in Fig. 6.3 at 90 min show that crystal growth was also affected by sonication. In addition to the induction of crystallization smaller crystals were obtained and different morphologies were observed. Crystals with a

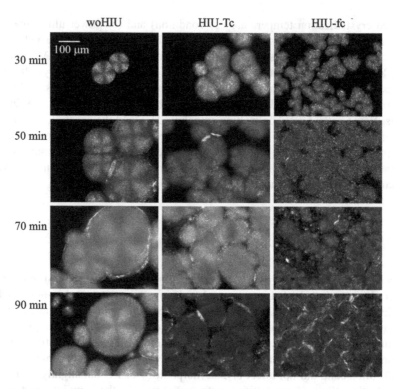

Fig. 6.3 Effect of power ultrasound on the crystal morphology of PKO as a function of crystallization time and application. The *first column* shows crystals obtained when PKO was crystallized without power ultrasound (woHIU), the *second column* shows crystals of PKO samples crystallized with power ultrasound applied under T_c condition (HIU-T_c), and the *third column* shows pictures of PKO crystals obtained after power ultrasound application under f_c condition (HIU-f_c). Reprinted from Suzuki et al. (2010) with kind permission from John Wiley and Sons

defined Maltese cross shape were observed in the PKO samples crystallized without power ultrasound and were not observed in sonicated samples. Previous research performed in cocoa butter (Higaki et al. 2001; Ueno et al. 2003) indicates that power ultrasound can change polymorphism of fats. The effect of power ultrasound on generating different polymorphic forms will be discussed in Sect. 6.5. It is likely that the two types of crystals observed for PKO correspond to different polymorphic forms in PKO. It is also evident from Fig. 6.3 that the effect of power ultrasound on PKO crystallization was more pronounced when applied as soon as the crystals are observed by the naked eye (i.e., f_c conditions).

A crystal network with smaller crystals was obtained in sonicated PKO samples which resulted in harder materials similar to the behavior observed in AMF. However, a significantly ($\alpha = 0.05$) harder texture was obtained only for the PKO sonicated under the f_c condition. These results suggest that the crystal size reduction obtained for the T_c condition was not sufficient to generate a significant increase in samples' hardness. It is surprising however that melting enthalpies of the sonicated samples were significantly higher ($\alpha = 0.05$) than non-sonicated samples. It is probable that the use of power ultrasound induced polymorphic transformations in PKO, as also suggested by the different morphologies observed in Fig. 6.3 resulting in different enthalpy and hardness values.

6.3 All-Purpose Shortenings

The effect of power ultrasound was also evaluated in two types of shortenings: (a) shortening with high content of saturated fatty acids (~ 50 %) and (b) shortening with low content of saturated fatty acids (~ 30 %). The objective of these experiments was to determine if power ultrasound can be used in low saturated fats to improve their physical properties including texture, melting behavior, and crystal morphology.

6.3.1 Experimental Conditions

All-purpose shortening with high content of saturated fatty acids: Suzuki et al. (2010) used power ultrasound in an all-purpose shortening composed of approximately 50 % saturated fatty acids (~ 38 % palmitic and 7 % stearic acids). The melting point of the sample was 35.1 ± 0.4 °C. Power ultrasound was applied as previously described for PKO (Sect. 6.2.1) using an acoustic power of 80 W and a crystallization temperature of 30 °C.

All-purpose shortening with low content of saturated fatty acids: Ye et al. (2011) used power ultrasound to induce the crystallization of a low saturated fat shortening. The shortening had approximately 30 % of saturated fatty acids of which 11 % was palmitic acid and 22 % was stearic acid. The melting point of this sample was 33.4 ± 0.5 °C. Samples were crystallized at different crystallization temperatures ($T_c = 26$, 28, 30, and 32 °C) and power ultrasound was applied under T_c and f_c conditions as previously described in Sect. 6.2.1. In addition, power ultrasound was applied at different acoustic power levels (44, 54, 61, 72, and 101 W). Samples were left in the crystallization cell for 90 min and then tempered at 25 or 5 °C for 48 h. Functional properties of these materials after 90 min in the crystallization cell and after tempering were measured using DSC, PLM, X-ray diffraction (XRD), TPA, and rheology (Ye et al. 2011).

6.3.2 Results and Discussion

All-purpose shortening with high content of saturated fatty acids: As expected, sonication induced the crystallization of the shortening generating smaller crystals as described for the AMF and PKO samples (Fig. 6.4). Similar to the results discussed for PKO, the hardness of the sonicated samples was only significantly ($\alpha = 0.05$) higher for samples sonicated under f_c conditions, while no significant differences were found between the melting enthalpies of the sonicated and non-sonicated shortenings.

Crystal sizes in these samples were smaller than those obtained for PKO and AMF due to the high supercooling used in the shortening. Although there was a significant reduction in the crystal size and a significant induction in the crystallization when the shortening was crystallized under the T_c condition, this effect is not translated into a difference in hardness or enthalpy values. However, these two

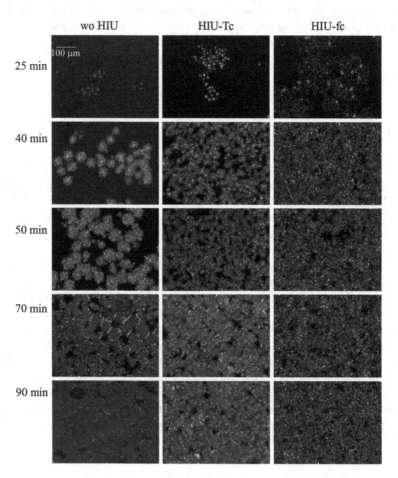

Fig. 6.4 Effect of power ultrasound on the crystal morphology of an all-purpose shortening as a function of crystallization time and power ultrasound application. The *first column* shows crystals obtained when the shortening is crystallized without power ultrasound (woHIU). The *second* and *third columns* correspond to crystals obtained with the shortening crystallized with power ultrasound under T_c (HIU-T_c) and f_c (HIU-f_c) conditions, respectively. Reprinted from Suzuki et al. (2010), with kind permission from John Wiley and Sons

sonication conditions resulted in a crystal network with a sharper melting profile as determined by DSC. This sharper melting profile was also observed for the sonicated AMF but was not observed for the sonicated PKO (Suzuki et al. 2010).

All-purpose shortening with low content of saturated fatty acids: Ye et al. (2011) show that the physical properties of a low saturated fatty acid shortening can be improved by using power ultrasound. These authors evaluated the effect of acoustic power levels on the crystallization behavior of the fat. Figure 6.5 shows pictures of the crystals obtained after 15 or 20 min in the crystallization process for samples crystallized with and without the use of power ultrasound. Columns in Fig. 6.5 show crystals obtained when samples were crystallized at different T_c; while rows in the figure show the effect of power levels (0, 44, 54, 61, 72 and 101 W) on fat crystallization. It is evident that at higher acoustic power used, there is more induction in crystallization and smaller crystals are formed for the same time point for all the crystallization temperatures tested. This effect of power ultrasound on inducing crystallization and on forming smaller crystals was also

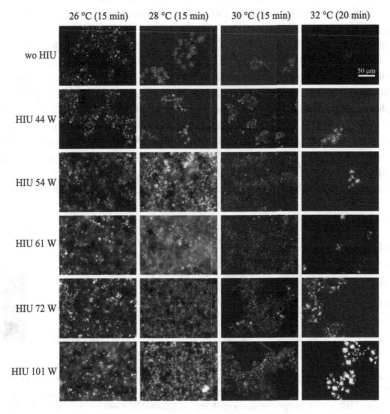

Fig. 6.5 Effect of power ultrasound on the crystal morphology of a low saturated shortening as a function acoustic power and crystallization temperature. Pictures were taken at 15 min for samples crystallized at 26–30 °C and at 20 min for samples crystallized at 32 °C. Reprinted from Ye et al. (2011) with kind permission from American Chemical Society, Copyright 2011

observed after the crystals were kept at T_c for 90 min (Ye et al. 2011). Crystals shown in Fig. 6.5 were obtained when power ultrasound was applied under T_c conditions as described in Sect. 6.2.1 (10 min into the crystallization process). Depending on crystallization temperature, power ultrasound induced either primary or secondary nucleation. When samples were crystallized at 26 and 28 °C some crystals were already present at the time of sonication (10 min into the crystallization process). Changes in crystallization behavior observed under these conditions suggest that power ultrasound induced secondary nucleation since it was applied in the presence of crystals. On the other hand, when samples were crystallized at high temperatures such as 30 and 32 °C, no crystals were observed at 10 min suggesting that power ultrasound induced primary nucleation under these conditions. To further evaluate the effect of sonication in the presence of crystals the shortening was crystallized at $T_c = 30$ and 32 °C and power ultrasound was applied when the first crystals (f_c condition) were observed using the highest acoustic power level tested (101 W). The f_c condition corresponded to an application time of 13 and 20 min for the samples crystallized at 30 and 32 °C, respectively. As expected, the effect of power ultrasound was enhanced when used under the f_c condition (Ye et al. 2011).

Changes in crystallization behavior induced by sonication led to crystal networks with different textures. All sonicated samples were harder than non-sonicated samples (Fig. 6.6); however, samples crystallized at 30 °C and sonicated under T_c conditions were not significantly harder than the non-sonicated samples. Only samples crystallized at this temperature using the f_c sonication conditions were significantly harder. Conversely, samples crystallized at 32 °C under f_c sonication conditions were significantly harder than samples crystallized and sonicated using T_c conditions which in turn were significantly harder than the non-sonicated samples. The effect of sonication in creating harder materials was so significant that similar hardness values were obtained at 32 °C in the sonicated samples

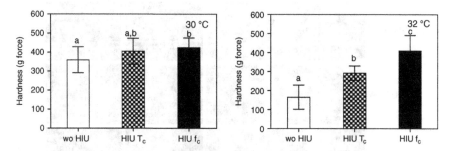

Fig. 6.6 Effect of sonication on the hardness of a low saturated shortening crystallized at 30 and 32 °C under different sonication conditions. "wo HIU": sample was crystallized without sonication, "HIU T_c": sample was crystallized and power ultrasound was applied when the sample reached crystallization temperature (10 min) and "HIU f_c": sample was crystallized and power ultrasound was applied when the first crystals were observed (13 and 20 min for the sample crystallized at 30 and 32 °C, respectively). Reprinted from Ye et al. (2011) with kind permission from American Chemical Society, Copyright 2011

(409.5 ± 81.4 g force) compared to the non-sonicated samples crystallized at 30 °C (359.3 ± 68.8 g force). These results clearly show that power ultrasound can be used in crystallization processes performed at higher temperatures without losing important functional properties of the material, such as hardness.

Changes in the microstructure and hardness of the crystal network generated by sonication were translated into changes in the viscoelastic behavior of these samples. The solid-like behavior of a crystalline network can be measured using the elastic or storage modulus (G′). Table 6.2 reports G′ values of sonicated and non-sonicated samples. In general, G′ values of sonicated samples were significantly higher ($\alpha = 0.05$) than those obtained in the non-sonicated samples. G' values increased as much as 22 times for samples crystallized at 32 °C, where G′ values of samples crystallized without power ultrasound were 195 ± 18 Pa and G′ values of samples crystallized with power ultrasound under f_c conditions were 4,229 ± 1,277 Pa. This effect of power ultrasound on the increase of the storage modulus (G′) was also observed after tempering the sample at 25 and 5 °C for 48 h. When samples crystallized at 32 °C with and without the application of power ultrasound under f_c conditions were tempered for 48 h at 5 °C the G′ values obtained were $1.2 \times 10^5 \pm 0.5 \times 10^5$ Pa and $9.7 \times 10^5 \pm 0.3 \times 10^5$ Pa, respectively. When the same samples were tempered for 48 h at 25 °C, G′ values were 25 ± 11 Pa and 1,719 ± 186 Pa for the samples crystallized at 32 °C without and with power ultrasound under f_c condition, respectively.

No significant differences were reported by Ye et al. (2011) in the melting onset temperature (T_{on}), melting peak temperature (T_p), or melting enthalpy (ΔH) values of samples crystallized with and without the application of power ultrasound. However, they did report a slight fractionation in melting profiles of these samples

Table 6.2 Storage modulus (G′) of sonicated and non-sonicated IESBO crystallized at different T_c for 90 min and after tempering for 48 h at 5 and 25 °C

	G′ (Pa)	
Sonication condition	30 °C	32 °C
	90 min	
Wo HIU	2,610 ± 83[a]	195 ± 18[a]
T_c condition	1,410 ± 152[b]	597 ± 160[a]
f_c condition	2,793 ± 53[a]	4,229 ± 1,277[b]
	48 h at 5 °C	
Wo HIU	$5.9 \times 10^5 \pm 0.8$[a]	$1.2 \times 10^5 \pm 0.5$[a]
T_c condition	$8.9 \times 10^5 \pm 0.6$[a]	$7.8 \times 10^5 \pm 0.4$[b]
f_c condition	$7.0 \times 10^5 \pm 2.0$[a]	$9.7 \times 10^5 \pm 0.3$[c]
	48 h at 25 °C	
Wo HIU	58 ± 8[a]	25 ± 11[a]
T_c condition	203 ± 120[a]	254 ± 117[a]
f_c condition	129 ± 88[a]	1,719 ± 186[b]

Mean values and standard deviations are reported. Mean values within a column and storage condition with the same superscript letter are not significantly different ($\alpha = 0.05$). Reprinted from Ye et al. (2011) with kind permission from American Chemical Society, Copyright 2011

with an evident shoulder at lower temperatures obtained for samples crystallized with power ultrasound. Finally, samples crystallized at 30 and 32 °C with the application of power ultrasound had a sharper melting profile compared to the same samples crystallized without power ultrasound. These results suggest that some type of molecular reorganization takes place during sonication that causes a partial fractionation in the sample during melting and sharper melting profiles.

6.4 Pure Triacylglycerols

Sonocrystallization of pure triacylglycerols was studied by Sato's group (Higaki et al. 2001). These authors evaluated the effect of sonication on the microstructure and polymorphism of tripalmitin and trilaurin using small angle and wide angle X-ray diffraction. Below is a summary of the experimental conditions used and the results reported.

6.4.1 Experimental Conditions

Tripalmitin (PPP): Tripalmitoilglycerol (PPP) was crystallized at different temperatures (46–54 °C) using a crystallization cell similar to the one described in Fig. 6.1. A 20 kHz acoustic wave was used at a power level of 100 W (Generator model DG-100-20; Telsonic Co., Bronschhoften, Switzerland). Power ultrasound was applied for 5, 10, 15, 30, 45, and 60 s (Higaki et al. 2001) and samples were crystallized under two conditions: (a) *isothermal conditions*, where the sample was sonicated at a specific temperature and then kept at this temperature for the duration of the experiment, and (b) *cooling crystallization*, where the sample was cooled at different cooling rates (1, 2, and 4 °C/min) after sonication. The progress of crystallization was followed using polarized light microscopy and the type of polymorphic form obtained was analyzed using X-ray diffraction (Rigaku, Tokyo, Japan).

Trilaurin (LLL): Ueno et al. (2003) studied the crystallization of LLL with and without the use of power ultrasound using SAXS and WAXS X-ray diffraction. Sonication was applied for 2 s using an acoustic wave of 20 kHz and an acoustic power of 100 W. LLL was crystallized at 30 and 25 °C and the induction time of formation of β' and β polymorphs were measured using the SAXS X-ray diffraction measurements.

6.4.2 Results and Discussion

Tripalmitin (PPP): Data reported by Higaki et al. (2001) show several interesting results. These authors observed a significant increase in the sample temperature

Table 6.3 Type of polymorphic form obtained in PPP crystallized at different temperature (T_c) with and without sonication (Higaki et al. 2001)

T_c (°C)	Without sonication	With sonication
46	β'	β'
48	β'	β'
50	β'	β'
52	β	β'
53	β	$\beta' + \beta$
54	β	$\beta' + \beta$

during sonication for 10 s under isothermal conditions (53 °C). The temperature in the sample increased from 53 to 80 °C and β and β' polymorphs were obtained. This significant increase in temperature was not observed in the studies performed by Martini's group who observed temperature increases of the order of 0.5–1 °C during sonication for 10 s. Higaki et al. used a power density of 50 W/cm^3, while Martini's group used power densities of the order of 1 W/cm^3. This difference in power density might be responsible for the temperature increase observed in Higaki's work. Table 6.3 summarizes the type of polymorphic forms obtained when PPP was crystallized at different temperatures with and without sonication.

When PPP was crystallized without sonication, β polymorphs were obtained at temperatures between 52 and 54 °C, while the β' polymorph was obtained at lower temperatures (46–50 °C). When samples were crystallized in the presence of sonication an induction of crystallization was observed for both polymorphic forms, especially at lower temperatures. In addition, the formation of β' crystals was observed at higher temperatures (52–54 °C) while β crystals were only observed at temperatures of 53 and 54 °C. These results indicate that power ultrasound inhibits the formation of β crystals and promotes the formation of β' polymorphism. In addition, when sonocrystallization experiments were performed with a low purity PPP, power ultrasound increased the proportion of β' crystals with respect to the amount of β crystals when crystallized at 52 °C and sonicated for 15 s. When PPP was sonicated at 54 °C for different periods of times and then cooled at 2 °C/min, the α polymorph was the predominant form at both short and long sonication times. Interestingly, the β' form was the most abundant when sonication was applied between 5 and 30 s with a maximum of β' content observed at 15 s.

Finally, Ueno et al. (2003) showed that when PPP was crystallized with and without sonication for 15 s the proportion of β' crystals decreased as the cooling rate increased. The same group of researchers performed sonocrystallization experiments in PPP where the sample was crystallized at 50 °C and power ultrasound (20 kHz) was applied using 100 W of power for 2 s. In these experiments 5 mL of sample was used with a power density of 20 W/cm^3. This lower power density resulted in a temperature increase of only 2.5 °C. As expected, an induction in PPP crystallization was observed when power ultrasound was applied. When PPP was crystallized without the use of power ultrasound, both β' and β polymorphic forms were obtained while only β crystals were with sonication.

Trilaurin (LLL): As previously described for PPP, Ueno et al. (2003) showed that the induction time of crystallization for the β' and β polymorphs was significantly reduced when power ultrasound was used for 2 s (100 W, 20 kHz) in LLL crystallized at 25 and 30 °C. A greater reduction in the induction time was observed for higher temperatures (30 °C) where induction times for the formation of β crystals were reduced from 2,820 to 30 s when LLL was crystallized without and with sonication, respectively. In addition, β' polymorphism was not obtained when LLL was crystallized with power ultrasound. When LLL was crystallized at 25 °C both β' and β polymorphs were formed with induction times of 60 and 300 s when crystallized without power ultrasound and 40 and 100 s when crystallized with power ultrasound, respectively.

These results show that power ultrasound can be used to tailor the formation of specific polymorphic forms in a crystallizing fat. This is especially important in processing highly polymorphic fats such as cocoa butter where the formation of a stable polymorph is needed to achieve the desired product quality and extend shelf life of the product.

6.5 Cocoa Butter

The effect of power ultrasound on the crystallization of cocoa butter was studied by Sato's group (Higaki et al. 2001) and was focused on polymorphic changes induced by high intensity acoustic waves.

6.5.1 Experimental Conditions

Power ultrasound was applied to cocoa butter (CB) from the Ivory Coast using a 20 kHz acoustic wave and a power of 300 W in a 250 mL sample. The sonication temperature was kept at 32.5 °C during 3 and 15 s of sonication and then samples were stored at 20 °C for 30 min followed by crystallization at 4 °C (Higaki et al. 2001). The polymorphic forms of the cocoa butter were analyzed using X-ray diffraction (Rigaku, Tokyo, Japan).

6.5.2 Results and Discussion

Polymorphic form II was obtained when CB was crystallized without power ultrasound, while the polymorphic form V was obtained when ultrasound was applied during crystallization for 3 s. When longer sonication times were applied (over 9 s) to CB a mixture of forms II and V was obtained but only form II was observed for very long sonication times (over 15 s). Higaki et al. (2001) explain these results by the increase in temperature generated during longer sonication times that resulted in the melting of form V.

These results confirm that power ultrasound can be used in cocoa butter for chocolate production to induce the formation of a desired polymorphic form. This suggests that power ultrasound may be a useful tool to delay blooming and to extend the shelf life of chocolate. Additional work will be required to optimize process parameters in the chocolate matrix.

6.6 Bubble Formation in Lipid Samples

As described in Chap. 5 cavitation is responsible for changes in crystallization behavior caused by acoustic waves. The formation and dynamics of the bubbles formed during sonication in lipids must be evaluated to further understand mechanisms of sonocrystallization. The presence of bubbles can be detected through optical means such as photography, light and electron microscopy, and light scattering. In addition, the detection of events created from bubble excitation during sonication can be quantified using cavitation erosion and sonochemistry techniques, while acoustic techniques (low intensity, high frequency ultrasound) can be used to detect both oscillating (or excited) and non-oscillating bubbles. Acoustic techniques have the advantage that they can be used with opaque materials and that they can be used to detect bubbles that oscillate at small amplitudes (Leighton 1994).

Martini et al. (2012) has recently studied the formation and life cycle of bubbles generated during sonication in an edible lipid (soybean oil) using low intensity ultrasound. The experimental conditions and results obtained will be discussed below.

6.6.1 Experimental Conditions

The formation of bubbles in soybean oil (SBO) during power ultrasound application was monitored using a low intensity ultrasound spectrometer (SAI-7, VN Instruments, ON, Canada) operating at a central frequency of 1 MHz. Power ultrasound was applied to SBO (100 mL) at different temperatures (22, 24, 26, 28, and 30 °C) for different periods of time (5, 10, and 60 s). A 20 kHz sonicator (Misonix 3000, Misonix Inc., NY) adapted with a 3.2 mm diameter tip was used for the sonication. Electrical powers of 6, 21, 42, 66, and 90 W were used to generate tip amplitude vibrations of 24, 72, 120, 168, and 216 μm.

6.6.2 Results and Discussion

Acoustic power levels obtained during these experiments were slightly affected by sonication time, sample temperature, and electrical power used. Acoustic power levels ranged from approximately 1 W for electrical power levels of 21 W to approximately 68 W for electrical power levels of 90 W.

The presence of bubbles in SBO was detected by measuring the attenuation of a low intensity acoustic signal as a function of frequency. Attenuation was caused by the scattering of bubbles in the media and the attenuation coefficient was measured in dB cm^{-1}. As expected, more bubbles were detected when SBO was sonicated at higher power levels. This effect was more significant at higher temperatures (30 °C) where attenuation values greater than 20 dB cm^{-1} were observed. In addition, more attenuation was observed toward the end of the sonication process, especially at lower frequencies, suggesting the presence of bigger bubbles. Rectified diffusion events during sonication might be responsible for bubble growth in the material and the presence of bigger bubbles toward the end of sonication (Chap. 5). In addition, some degree of attenuation was observed at low frequencies even after the sonication was stopped. This attenuation was of the order of 7–15 dB cm^{-1} and might be caused by bigger bubbles remaining in the media after sonication stopped. This effect was more evident at shorter sonication times and lower temperatures. The presence of bubbles after sonication ceased was also shown in reflection measurements. These results showed that the number of bubbles in the media after the sonication ceased depends on the duration of sonication and on sample temperature (Martini et al. 2012).

These are the first set of data that corroborate the presence of cavities or bubbles generated during sonication of lipids. Further research in this area is required to better understand the role of cavity formation and the life cycle of the cavities and how these cavities induce crystallization in lipids.

6.7 Other Effects to Consider

6.7.1 Effect of Power Ultrasound on Oxidation

Results presented in Sects. 6.1–6.6 discuss the use of power ultrasound to induce crystallization, induce or delay the formation of specific polymorphic forms, and generate smaller crystals and harder materials with sharper melting profiles. Acoustic waves, however, might have other effects on the material when applied under specific conditions. Chemat et al. (2004) described the generation of oxidation products in sonicated sunflower oil. These authors claim that oxidation products are induced or enhanced by the presence of free-radicals generated during sonication. In this study sunflower oil (100 ml) was sonicated between 0.5 and 30 min using a 20 kHz probe (13 mm diameter) and an electrical power of 150 W. These conditions are more aggressive than those used in sonocrystallization experiments previously reported (Sects. 6.1–6.6), where the maximum sonication time was 15 s. Recent research performed in our laboratory showed no effect of sonication on oxidation stability of soybean oil (SBO) and a low saturated

Fig. 6.7 Peroxide value
(PV) of sonicated (S) and
non-sonicated (NS) soybean
oil (SBO) and a low saturated
shortening (IESBO) stored at
25 °C for 190 days

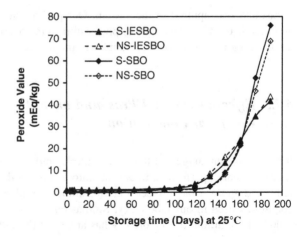

shortening (IESBO) as measured by peroxide value (Fig. 6.7). Samples were
sonicated for 10 s using 100 W of acoustic power and a 3.2 mm diameter tip.
Sonicated and non-sonicated samples were stored at 25 °C for 190 days in the dark
and in the presence of oxygen. The onset of oxidation for the IESBO was
approximately 120 days, while the onset of oxidation for the SBO was approximately 135 days. Even though the onset of oxidation was shorter for the IESBO,
PV of SBO samples were significantly higher than those observed for IESBO after
190 days of storage. It is important to note here that the PV of fresh refined oils is
usually below 1 mEq/Kg. Oils with PV between 1 and 5 mEq/Kg are considered
slightly oxidized, oils with PV between 5 and 10 mEq/Kg are considered mod-
erately oxidized, while oils with PV above 10 mEq/Kg are considered highly
oxidized (Gunstone 2008; O'Brien 2009). Results reported in Fig. 6.7 show no
differences in the PV of the sonicated and non-sonicated SBO or IESBO samples,
especially for PV values below 10 mEq/Kg. These results are in accordance with
data reported by Patrick et al. (2004) who evaluated the effect of power ultrasound
on generation of off-flavors. These authors sonicated sunflower oil for 5 min using
a 20 kHz probe and detected only very small quantities of oxidation products such
as benzene.

6.7.2 Effect of Agitation on Power Ultrasound Efficiency

This chapter introduced various processing parameters including cooling rate,
crystallization temperature, and sample volume that can affect the efficiency of
power ultrasound in inducing crystallization. An additional parameter that might
influence the efficiency of power ultrasound is the presence of agitation. In most
experiments performed by Martini's laboratory power ultrasound was applied
under static conditions. This means that no agitation was used after power

ultrasound was applied to the system. Our preliminary results suggest that when agitation is used the efficiency of power ultrasound decreases. This could be due to a rapid dissolution of acoustic bubbles in the media as a consequence of agitation.

6.7.3 Effect of Power Ultrasound on Fat Chemical Composition

It has also been suggested that power ultrasound can induce chemical changes in lipids. Research performed in our laboratory showed that there is no change in the fatty acid and/or triacylglycerol composition of lipid samples sonicated under the conditions used in our laboratory. Similar results were reported by Chemat et al. (2004). In addition, our laboratory has analyzed the fatty acid and triacylglycerol composition of crystals obtained from sonicated and non-sonicated samples. After sonication, crystals were isolated by filtration and their melting behavior and chemical composition analyzed. No significant differences were observed between the chemical composition of the sonicated and non-sonicated samples or in their melting behavior. These results suggest that no differential crystallization is observed as a consequence of sonication.

6.8 Conclusions

Power ultrasound can be used as an additional processing tool to induce the crystallization of lipids, to tailor the formation of specific polymorphic forms, and to generate small crystals which result in a harder material with sharper melting profiles. The efficiency of ultrasound has been demonstrated in several lipid systems such as anhydrous milk fat (Martini et al. 2008; Suzuki et al. 2010), palm kernel oil and high saturated shortenings (Suzuki et al. 2010), low saturated shortening (Ye et al. 2011), palm oil (Patrick et al. 2004), tripalmitin and trilaurin (Higaki et al. 2001; Ueno et al. 2003), and cocoa butter (Higaki et al. 2001).

The efficiency of power ultrasound in inducing crystallization depends on several processing parameters:

- Initial material (fatty acid and triacylglycerol composition)
- Crystallization temperature
- Cooling rate
- Agitation
- Type of crystallization (isothermal vs. non-isothermal)
- Time at which power ultrasound is applied.

In addition, sonication parameters can also affect the efficiency of acoustic waves in inducing crystallization. These sonication parameters include:

- Acoustic power
- Acoustic density and intensity
- Size of the tip.

It is hypothesized that sonocrystallization in fats is mediated by:

- Formation and collapse of cavitation bubbles that might act as nucleation sites and therefore induce primary nucleation.
- Generation of high shear forces or agitation to allow for better molecular mobility and therefore an induction of primary and secondary nucleation.
- Increase in localized pressures which result in a higher melting point of the sample and therefore a higher supercooling which ultimately induces crystallization.

The use of power ultrasound to modify crystallization behavior of fats is still in its infancy and much more research is needed to fully understand the interaction among acoustic waves, cavities, and crystals. The use of sonocrystallization must be explored in several other lipids to broaden the use of this technique for other fat-based products. Specific areas of research that this author thinks that need to be addressed will be discussed in Chap. 7.

References

Arends BJ, Blindt RA, Janssen J, Patrick M (2003) Crystallization process using ultrasound. US 6,630,185 B2

Baxter JF, Morris GJ, Gaim-Marsoner G (1997a) Process for accelerating the polymorphic transformation of edible fats using ultrasonication. EP 0 765 605 A1

Baxter JF, Morris GJ, Gaim-Marsoner G (1997b) Process for retarding fat bloom in fat-based confectionery masses. EP 0 765 606 A1

Bund RK, Pandit AB (2007a) Rapid lactose recovery from paneer whey using sonocrystallization: a process optimization. Chem Eng Proc 46:846–850

Bund RK, Pandit AB (2007b) Rapid lactose recovery from buffalo whey by use of "antisolvent" ethanol. J Food Eng 82:333–341

Chemat F, Frondin I, Costes P, Moutoussamy L, Shum Cheong Sing A, Smadja J (2004) High power ultrasound effects on lipid oxidation of refined sunflower oil. Ultrason Sonochem 11:281–285

Chow R, Blindt R, Chivers R, Povey M (2003) The sonocrystallisation of ice in sucrose solutions: primary and secondary nucleation. Ultrasonics 41:595–604

Chow R, Blindt R, Kamp A, Grocutt P, Chivers R (2004) The microscopic visualisation of the sonocrystallisation of ice using a novel ultrasonic cold stage. Ultrason Sonochem 11:245–250

Chow R, Blindt R, Chivers R, Povey M (2005) A study on the primary and secondary nucleation of ice by power ultrasound. Ultrasonics 43:227–230

Dhumal RS, Biradar SV, Paradkar AR, York P (2008) Ultrasound assisted engineering of lactose crystals. Pharm Res 25:2835–2844

Gunstone FD (2008) Analytical parameters. In: Gunstone FD (ed) Oils and fats in the food industry. Wiley Blackwell, West Sussex, pp 37–58

Higaki K, Ueno S, Koyano T, Sato K (2001) Effects of ultrasonic irradiation on crystallization behavior of tripalmitoylglycerol and cocoa butter. J Am Oil Chem Soc 78:513–518

Leighton TG (1994) Effects and mechanisms. In: Leighton, TG (ed) The acoustic bubble. Academic Press, New York, pp 439–590

Li B, Sun DW (2002) Effect of power ultrasound on freezing rate during immersion freezing. J Food Eng 55:277–282

Martini S, Suzuki AH, Hartel RW (2008) Effect of high intensity ultrasound on crystallization behavior of anhydrous milk fat. J Am Oil Chem Soc 85:621–628

Martini S, Tejeda-Pichardo R, Ye Y, Padilla SG, Shen FK, Doyle T (2012) Bubble and crystal formation in lipid systems during high-intensity insonation. J Am Oil Chem Soc 89:1921–1928

O'Brien RD (2009) Fats and oils analysis. In: O'Brien RD (ed) Fats and oils: formulating and processing for applications, 3rd edn. CRC Press, Boca Raton, pp 197–250

Patel SR, Murthy ZVP (2009) Ultrasound assisted crystallization for the recovery of lactose in an anti-solvent acetone. Cryst ResTechn 44:496–889

Patrick M, Blindt R, Janssen J (2004) The effect of ultrasonic intensity on the crystal structure of palm oil. Ultrason Sonochem 11:251–255

Sun DW, Li B (2003) Microstructural change of potato tissues frozen by ultrasound-assisted immersion freezing. J Food Eng 57:337–345

Suzuki A, Lee J, Padilla S, Martini S (2010) Altering functional properties of fats using power ultrasound. J Food Sci 75:E208–E214

Ueno S, Ristic RI, Higaki K, Sato K (2003) In Situ studies of ultrasound-stimulated fat crystallization using synchrotron radiation. J Phys Chem B 107:4927–4935

Ye Y, Wagh A, Martini S (2011) Using high intensity ultrasound as a tool to change the functional properties of interesterified soybean oil. J Agric Food Chem 59:10712–10722

Chapter 7
Future Trends

Ultrasound offers great potential as a processing tool to improve functional properties of lipids and increase production efficiency, cost effectiveness, and environmental sustainability for the industry. However, research in the sono-crystallization of lipids is only at its infancy and much is left to be explored.

Notwithstanding evidence that power ultrasound affects crystal growth, additional research is required to identify factors responsible for this effect. It is not clear for example, the role that cavities or bubbles play in the growth of crystals during sonication. In addition, little is known about the effects that sonication has on the nucleation rate of fats and on the activation free energies needed to form new nuclei when using power ultrasound. There is a need to study bubble dynamics associated with sonocrystallization of fats and further understand the phenomena at the molecular level. The role that inertial and non-inertial bubbles play in inducing crystallization must be established by measuring cavitation thresholds. In addition, minimum power levels needed to induce crystallization using acoustic waves must be determined. Only one paper has addressed this need (Patrick et al. 2004) and further research is required to understand the physical phenomenon responsible for sonocrystallization in fats and expand this knowledge to other lipid systems.

The effect of power ultrasound on polymorphism requires further investigation for other fats such as palm oil, palm kernel oil, and cocoa butter that are commonly used by the food industry. The quality of lipid-based foods is often related to polymorphism present in the fat. The effect of ultrasound on generating different polymorphic forms depends on processing conditions such as cooling rate and crystallization temperature. Therefore, both sonication and other process parameters must be optimized to consistently generate the required polymorphic form for a specific lipid-based food product.

Fats are complex systems composed of a combination of triacylglycerols that interact with each other to form a solid material when cooled. Molecular interactions between these entities are driven by processing conditions such as temperature, cooling rate, and agitation, to name a few and are ultimately responsible for the macroscopic characteristics of the material formed after

S. Martini, *Sonocrystallization of Fats*, SpringerBriefs in Food, Health, and Nutrition, DOI: 10.1007/978-1-4614-7693-1_7, © Silvana Martini 2013

crystallization. There is very little known about the effect of fat chemical composition on the efficiency of power ultrasound in inducing crystallization. To better predict the effect of power ultrasound on fat crystallization we need to understand what role chemical composition plays in sonocrystallization events. For example, sonication of fats with different type and amount of saturated fatty acids and triacylglycerol species might result in crystalline networks with different characteristics. Given the research published to date, it is evident that the chemical composition of the starting material determines sonication conditions needed to induce crystallization.

The long-term effects of using power ultrasound to influence product quality are yet to be established. No data have been accumulated on the physical and sensorial stability of sonicated fat. For example, the preservation of a specific texture or polymorphic form obtained by sonication as a function of storage conditions such as time and temperature need to be established. Similarly, generation of off-flavors needs to be explored in sonicated fats stored for long periods of time. Food products must be prepared using sonicated fats and their quality quantified in terms of product performance, consumer acceptability, and shelf life.

There are certainly many challenges ahead for food scientists interested in exploring various mechanisms by which sonocrystallized fats are formed and to extend this knowledge to produce foods with improved nutritional properties and optimized quality.

Reference

Patrick M, Blindt R, Janssen J (2004) The effect of ultrasonic intensity on the crystal structure of palm oil. Ultrason Sonochem 11:251–255

Index

A

Acoustic
  frequency, 8, 23, 35, 38
  intensity, 17, 19, 21, 38
  power, 17, 19–22, 30, 35, 36, 49, 51, 52, 57, 59
  wave, 4, 7–12, 17–19, 21, 22, 27, 30, 35, 41, 45, 54, 58, 63
Anhydrous milk fat, 41, 60
Attenuation, 17–19, 58

B

Bubble, 36, 37, 57, 58, 63

C

Cavitation
  inertial, 36, 45
  non-inertial, 36, 45
Cocoa butter, 1, 3, 48, 56, 57, 60, 63
Crystallization temperature, 3, 38, 42, 43, 47, 49, 51, 52, 59

D

Defoaming, 27, 28
Diagnostic ultrasound, 8, 11

E

Emulsification, 20, 27, 28
Extraction, 27

F

Fats, 1–4, 9, 11, 41, 45, 49, 56, 61, 63, 64

H

Hardness, 42, 45, 47, 49, 50, 52, 53

L

Low intensity ultrasound, 18, 57

M

Melting behavior, 1, 2, 45, 49, 60
Microstructure, 53, 54

O

Oxidation, 58, 59

P

Palm kernel oil, 46, 63
Palm oil, 46, 60, 63
Pasteurization, 29
Peroxide value, 59
Power ultrasound, 4, 12, 19, 22, 27–30, 37, 38, 41, 43–52, 54, 56–59

R

Rectified diffusion, 36, 37

S. Martini, *Sonocrystallization of Fats*, SpringerBriefs in Food, Health, and Nutrition, DOI: 10.1007/978-1-4614-7693-1, © Silvana Martini 2013